T0339893

Vehicle Collision Dynamics

Analysis and Reconstruction

Vehicle Collision Dynamics

Analysis and Reconstruction

Dario Vangi
Department of Industrial Engineering, University of Florence,
Florence, Italy

Butterworth-Heinemann is an imprint of Elsevier
The Boulevard, Langford Lane, Kidlington, Oxford OX5 1GB, United Kingdom
50 Hampshire Street, 5th Floor, Cambridge, MA 02139, United States

Notices
Knowledge and best practice in this field are constantly changing. As new research and experience broaden our understanding, changes in research methods, professional practices, or medical treatment may become necessary.

Practitioners and researchers must always rely on their own experience and knowledge in evaluating and using any information, methods, compounds, or experiments described herein. In using such information or methods they should be mindful of their own safety and the safety of others, including parties for whom they have a professional responsibility.

To the fullest extent of the law, neither the Publisher nor the authors, contributors, or editors, assume any liability for any injury and/or damage to persons or property as a matter of products liability, negligence or otherwise, or from any use or operation of any methods, products, instructions, or ideas contained in the material herein.

British Library Cataloguing-in-Publication Data
A catalogue record for this book is available from the British Library

Library of Congress Cataloging-in-Publication Data
A catalog record for this book is available from the Library of Congress

ISBN: 978-0-12-812750-6

For Information on all Butterworth-Heinemann publications
visit our website at https://www.elsevier.com/books-and-journals

Publisher: Mathew Deans
Acquisition Editor: Carrie Bolger
Editorial Project Manager: Joshua Mearns
Production Project Manager: Selvaraj Raviraj
Cover Designer: Christian J. Bilbow

Typeset by MPS Limited, Chennai, India

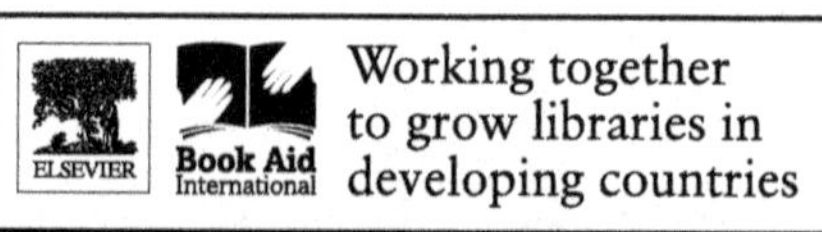

Dedication

I dedicate this book to my sons and my parents

Contents

Preface

Collision dynamics concerns the evolution of physical, kinematic, and dynamic quantities, during the impact between a vehicle and another obstacle, vehicle, or element of the infrastructure. The analysis of the collisions dynamic is mainly involved in the field of vehicle safety. It can be carried out in different ways, depending on the purpose and objectives and can only concern vehicles or even passive protection and restraint systems. The purposes of the analysis can be schematically summarized in (1) predictive analysis, in the field of vehicle design, generally conducted through numerical simulations; (2) verification of the safety and crashworthiness of the vehicle, as part of the tests on prototypes in the vehicle development phase and in the approval or consumer test phase (EuroNCAP type); (3) analysis of real accidents, as part of the study on the causes and modalities of accidents, both for road safety and in the judiciary field, conducted through the reconstruction of the event.

Methods used for the collision dynamics analysis are typically based (1) on the application of finite-element method (FEM) or multibody system (MBS) models and/or with the use of analytical or numerical models; (2) on the analysis of the signals acquired during the tests, both on the vehicle and the dummies; and (3) on the use of analytical or numerical models.

The FEM involves the discretization of the vehicle in a large number of elements, on which the governing equations are applied, generally nonlinear. The method is accurate, but time-consuming and is used in cases where accuracy balances the computational cost, for example, in the analysis of the structural response of newly designed vehicles, relative to the entire vehicle or to individual components.

When a detailed analysis of the deformation and stresses acting on the individual components or parts of the vehicle is not required, a modeling of the MBS is often used, in which the vehicle is discretized in different rigid parts, mutually connected through kinematic joints. The forces are exchanged through these constraints, and the equations of motion are obtained with the application of the Lagrangian method of dynamics. In quite a number of software, it is possible to apply additional FEM modules to take into account the deformation of the individual parts. The MBS method generally allows a faster analysis than the FEM; it can be used in the early stages of vehicle design to analyze the stresses on various components and occupants during an impact and for the simulation of advanced driver-assistance systems.

Impulsive models are essentially based on Newton's second law, the principle of impulse and momentum, and the principle of work and energy. These models make it possible to determine the deformation energy of the vehicles and the velocities after the impact starting from the initial conditions (forward reconstruction) or vice versa (backward reconstruction). This method is widely used due to the low calculation times but does not provide complete information on vehicle deformations or accelerations in the event of an accident.

A different approach is based on response surface models (RSM), used in accident reconstruction and crashworthiness analyses. The vehicle impact behavior is determined through the use of a testing campaign considering all intended parameters, and the acquired data are fitted to generate an analytical formulation describing the vehicle behavior. The vehicle features are thus reconstructed making use of calculations, but no commercially special-purpose software is available to automate the process.

Among all different available approaches, reduced order dynamic models (RODM) are typically employed when intermediate calculation time and accuracy are required; modeling is based on a simplification of methods previously listed. The most employed category of RODM is the lumped mass model that substitutes masses, dampers, and springs to structural elements.

While pure MBS models can accurately simulate the impact kinematics, no deformation is computed. Regardless of the application field, this lack of information is an important disadvantage in the model-validation process, since compatibility with real deformed shapes is an efficient strategy to assess the correctness of a reconstructed event. Although FEM and RSM model approaches calculate the deformations, they require considerably high computational resources and therefore take long simulation time. Also, from a practical point of view, limited availability of vehicle models and complexity in the ex novo modeling process of a vehicle makes these approaches less suitable for the analysis of real accidents, in which virtually any vehicle model/make can be involved. In these cases, the use of impulsive models may be the most appropriate.

This book aims to make clear the basic principles of vehicle impact dynamics and to give the reader an overview on the actual techniques, physical and mathematical, of applying the most common models used to analyze the vehicle impacts. These models can be used either in a simulation of possible impacts, sometimes used to finalize a subsequent FEM or multibody analysis, or in a reconstruction of a real accident. The material covered also lays the foundation for a later study and application of numerical models, such as FEM and MBS.

The method of presentation, as well as the examples, has been developed over almost 20years while teaching vehicle accident reconstruction to students and engineers at the University of Florence postgraduate and master courses.

The book is directed to seniors and first-year graduate students of physics and engineering; to those practicing scientists and engineers who wish to become familiar with the vehicle impact dynamic; and to all technicians who deal with vehicle safety or road safety, in general, accident analysis or court experts.

The text is divided into five chapters. The first is an introductory chapter where, after a brief panoramic of the main feature of vehicle structures focusing on crashworthiness, the phenomenological aspects of the crash are presented. The acceleration/time, velocity/time, and deformation/time curves acquired during a generic crash test are reported. The correlation between these curves and the vehicle structures involved in the crash process are well represented by the main characteristic of the force–deformation curves. The main parameters influencing the force–deformation curves are also discussed, as the test speed, the crash configuration (offset, underride, etc.).

In Chapter 2, Impact impulsive models, the impulsive model of impact between two vehicles is presented, based on the rigid body schematization of the vehicles with three degrees of freedom in the plane. The fundamental concepts to analyze the impact between two vehicles, namely the impact plane, the center of impact, the coefficient of restitution, and the energy loss in vehicle deformation are introduced. The momentum and energy conservation are then discussed, deriving the equations to analyze the impact both in backward and in forward mode. In particular, equations for unknown speeds are obtained knowing postimpact or preimpact speeds and the energy loss. Finally, the equations that provide the velocity changes and accelerations for different points of the vehicle are obtained.

The impulsive models described in Chapter 2, Impact impulsive models, allow evaluating the kinematic and energy parameters of an impact. However, in the impulsive models, the time variable t does not explicitly appear, because the collision is considered an instantaneous event, without duration: only the instants at the beginning and the end of the collision are considered. To describe what happens during the collision, for example, to describe the trend of accelerations, and crushing of vehicles, the time variable must be entered. In Chapter 3, Models for the structural vehicle behavior, the most widespread and usable models for the reconstruction of road accidents are examined, which describe the behavior of vehicles during the impact. Some models describe the vehicle's structural behavior by schematizing the vehicle as a lumped mass system, from which the differential equations governing its dynamics (lumped mass models) can be derived, or through the direct approximation of acceleration over time with analytical functions (pulse models). Finally, a method of direct integration of the motion equations of the vehicles and a model based on a discretization of the vehicle boundary (external perimeter) into elements (reduced-order discretization—RODM) is illustrated.

Chapter 4, Energy loss, deals with the problem of energy-loss assessment after a collision. More specifically, three classical approaches to estimate the energy loss are presented, one based on the classical spring-mass vehicle approximation, one on the more empirical energy equivalent speed parameter, and finally, a third method is called the Triangle method, which combines some of the features of the previous methods.

Chapter 5, Crash analysis and reconstruction, explains the criteria and the steps necessary to analyze and reconstruct the impact phase between two vehicles. The procedures to apply impulsive models, based on the conservation of momentum and angular momentum, and to apply models based on the relationships between force and deformation of vehicles, are analyzed. Finally, an example is illustrated, in which all the steps to evaluate the kinematic parameters and the reconstruction of the impact phase are explained for a real case.

In the text the following conventions are used: the vector "o" matrix quantities are indicated in bold, the vector product is indicated by the symbol "∧", and the scalar product with the symbol "•".

Acknowledgment

I owe a particular debt of gratitude to Prof. Sergio Reale for his advice, encouragement, and guide at various stages of my educational career.

I wish to express my gratitude to my collaborators and in particular I thank Antonio Virga, Filippo Begani, Carlo Cialdai e Michelangelo Gulino, as coauthors of publications, whose research results were used in this book.

Chapter 1

Structural behavior of the vehicle during the impact

Chapter Outline

1.1 Crashworthiness structures and phenomenological aspects of the impact

The bodywork must offer the necessary resistance to static and dynamic stresses induced during the motion of the vehicle, to ensure an adequate flexural and torsional stiffness and to protect the occupants of the car in case of an accident.

The type of bodywork used by most of the vehicles in production is the monocoque (uni-body) of molded steel sheet (Fenton, 1999; Heisler, 2002). The bearing shell is constituted by a spatial structure composed of more subtle elements of complex shape, joined together through spot welding. These elements contribute through their stiffness to the structural behavior of the body. The floor pan is constituted by the longitudinal members, the plane sheets, the crosspieces, the wheel arches, and the eventual transmission tunnel. The side pillars and the upper pavilion are fixed on the floor pan; the pillars are named A, B, and C pillars, which define the shape of the bodywork.

In the event of a frontal or lateral collision, different structures of the monocoque are involved, as shown in Fig. 1.1.

In some vehicles, typically commercial vehicles, off-road vehicles, and some Sport Utility Vehicles (SUV) and Multi-Purpose Vehicles (MPVs),

Vehicle Collision Dynamics. DOI: https://doi.org/10.1016/B978-0-12-812750-6.00001-9

the body is made with a separate chassis, realized through longitudinal members (body on frame or ladder chassis). This type of chassis is constituted by a substantially planar structure, composed of two longitudinal elements (spars), connected by several transverse elements (crossbars), which is entrusted with the task of providing resistance to lateral forces and to conferring torsional stiffness (see Fig. 1.2).

Recently, the construction of the body concerning the "space frame" (SF) geometry has been widespread, particularly indicated to be made of aluminum alloys. The SF is formed by a reticular structure consisting of a network of elements connected at the ends through rigid joints, typically made by casting, to form spatial geometries. Support functions of both the power train and chassis components are assigned to this spatial cage.

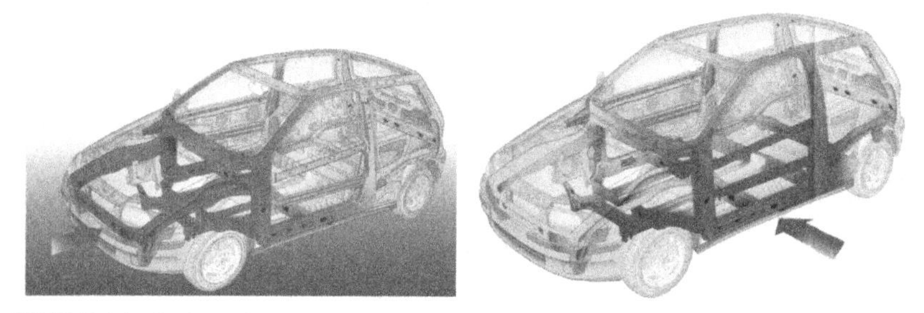

FIGURE 1.1 Bodywork structure of the vehicle, with the highlighted parts that most absorb the impact energy in case of a frontal (left) and lateral impact (right).

FIGURE 1.2 Example of a frame consisting of two longitudinal members, on which the bodywork is mounted.

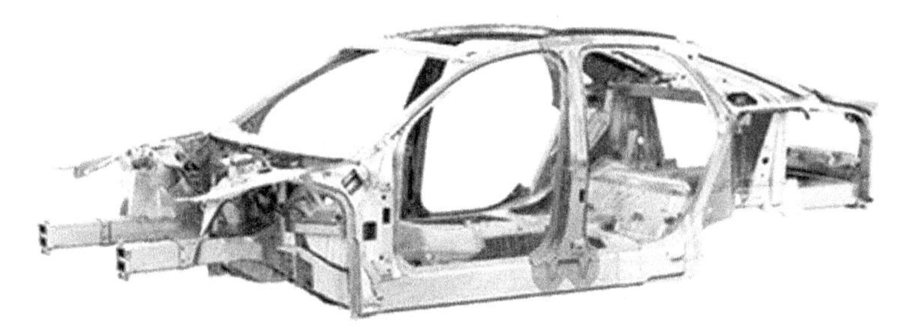

FIGURE 1.3 Example of a vehicle body, realized with space frame geometry.

Moreover, structural tasks to confer adequate stiffness to the vehicle and protect passengers in case of an accident are assigned to the spatial cage (Fig. 1.3).

The core idea of crashworthiness structure design is to preset a crumple zone, which can absorb the kinetic energy of vehicles during crashes, possibly lowering the acceleration. In a frontal crash, for example, the stiffness of the front structure determines the acceleration pulse during a crash. This pulse should have a specific shape, to minimize the risk for the occupant. During a massive collision, there are three essential phases (Witteman, 1999) as follows:

1. Crash initiation phase. In this phase the sensor triggering for the belt pretensioner and the airbag must take place. For optimal sensor triggering the front end of the car should be sufficiently stiff to generate within a short time interval a velocity change that lies above the trigger value of about 6 km/h.
2. Airbag deployment phase. In this phase the airbag is inflated, and the occupant tightens the belts while moving forward with a relative velocity with respect to the car. To minimize the injuries due to the impact with the airbag, the deceleration of the car should be sufficiently low in this phase, implying that the stiffness must be relatively low.
3. Occupant contact phase. In this phase the occupant has hit the airbag and there is a stiff contact between the occupant and the car. In this phase, high decelerations may occur because the occupant will not be subjected to further shock loads caused by contacts with the interior. The frontal car structure should be stiff enough to decelerate substantially in the remaining time.

Research (Brantman, 1991) has shown that for optimal occupant safety in a collision at 48 km/h impact velocity, the first phase lasts between 10 and 30 ms, the second phase lasts 35 ms, and the last phase fills up the remaining time to a total of maximal 90 ms.

The design process of the crashworthiness structure seeks to control the paths of load transformation and to optimize the energy-absorbing process

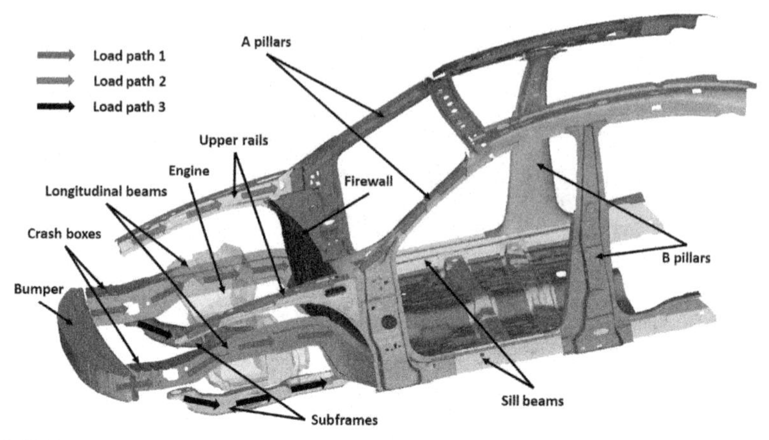

FIGURE 1.4 Load paths of vehicle frontal structure.

and acceleration pulse. For example, there are several load paths in the frontal structure of a modern Sedan, as shown in Fig. 1.4:

Path 1: Accessories—bumper—crash boxes—longitudinal beams
Path 2: Upper rails—A pillar
Path 3: Subframes—sill beams

The components in three paths are deformable and can absorb the impact energy. Notably, the first path absorbs more than 50% of the total crash energy in most frontal crashes (Griškevicius and Žiliukas, 2003).

Fig. 1.5 shows a rough estimation of energy absorption distributed on the different components of the frontal structure during a crash at 56 km/h against a rigid barrier.

The features and functions of these components are as follows:

- Bumper: the bumpers are usually reinforcement bars made of steel, aluminum, plastic, or composite material and can absorb crash energy to a certain extent. The primary purpose of a bumper is to minimize the cost of repair after low-speed crashes. It can also benefit the protection of pedestrians.
- Crash boxes: the crash boxes are generally thin-walled tubes with well-designed cross-sectional shape and crumple points (e.g., ditches and crash beads). They may collapse in a particular pattern to absorb energy efficiently.
- Longitudinal beams: the longitudinal beams are also a thin-walled structure, but longer and stronger than crash boxes. The deformation modes of longitudinal beams include folding, tearing, and bending.

Some reinforcing components may be used to strengthen the beams and optimize the energy absorbing.

In most cases, the deformations of these components in frontal impact do not occur at the same time, but in the order from front to rear. In fact, in each section of the vehicle, the forces must balance the inertia of the vehicle rear parts decelerated during the impact (see Fig. 1.6). So, the most significant forces are generally near the point of contact with the obstacle because here we have the inertia of the entire vehicle.

During impact, the most stressed structures are subjected to buckling (Jones, 2012; Stronge, 2000), with the plastic collapse of the various elements. Such plasticized zones extend away from the impact zone to more remote areas when the impact phase proceeds. Macroscopically, a wave or plastic flow advances from the area closest to the obstacle, affecting

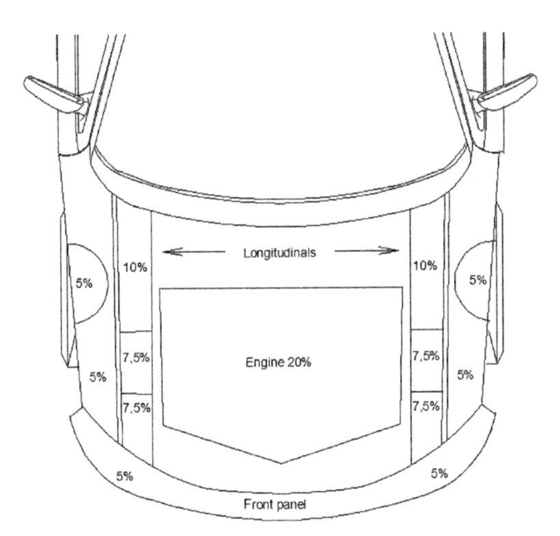

FIGURE 1.5 Estimated energy absorption percentages in the frontal structure. *From Witteman, W.J., 1999. Improved Vehicle Crashworthiness Design by Control of the Energy Absorption for Different Collision Situations. Technische Universiteit Eindhoven, Eindhoven. doi:https://doi.org/10.6100/IR518429.*

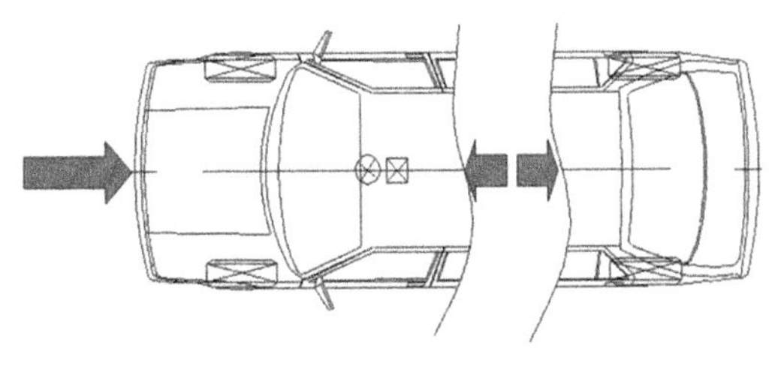

FIGURE 1.6 In a frontal collision, the most significant force occurs in the front, due to the inertia of the entire mass of the vehicle, while in a middle section, it is only acting the inertia of the rear part of the vehicle.

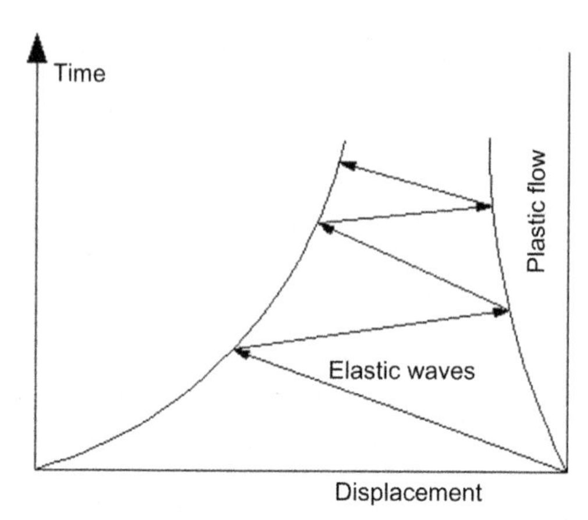

FIGURE 1.7 Plastic flow and elastic waves propagating in the vehicle structure as a result of *an impact.*

gradually more distant areas, with a propagation speed given by the closing motion between the vehicle and the obstacle. Together with the plastic flow, there is the propagation of elastic waves inside of the vehicle structure, which propagated faster, at the sound speed in the material of the structure. These waves are subjected to many reflected paths back and forth, affecting the whole structure, giving rise to the vibrations of the entire structure. The whole process is depicted schematically in Fig. 1.7.

Besides the deformable part, some components in the frontal vehicle structure should be strong enough. In most crashworthiness studies, engine and firewall are generally considered rigid bodies. Especially, the firewall refers to the rigid wall between the engine room and passenger cabin. If a vehicle crashes, the firewall can prevent the intrusion of vehicle cabin and, therefore, ensure enough living space for driver and passengers.

1.2 Pulse acceleration curve

Consider a crash test, like the one shown in Fig. 1.8, in which a vehicle collides frontally against a rigid, flat, immovable barrier. The frontal area of the vehicle, interested by the plastic flow, undergoes deformation due to the direct contact with the barrier and an induced deformation. The latter can extend beyond the area of direct contact, sometimes up to noncontiguous areas, far from those affected by direct contact. The induced deformation is caused both by the deformations of structures in direct contact with the barrier and by inertial loads that affect, to some degrees, a large part of the vehicle. Observing Fig. 1.8, we note a crease on the pavilion, well away from the area of direct contact.

The decelerations of the components inside the plastic flow area are different from those of the remaining parts of the vehicle. In fact, in the plastic

FIGURE 1.8 Example of crash test against a rigid barrier, performed at the LaSIS (Road safety and accident analysis laboratory)—University of Florence.

flow area, the components undergo a sharp deceleration and gradually stop while the remaining parts of the vehicle are still approaching the barrier. By placing accelerometers in different vehicle positions, different decelerations/time curves are expected. In addition, it must be considered that the entire structure is subject to the propagation of elastic waves during the impact time and then, at various points, different oscillations of deceleration due to the vibrations of the structures are expected.

Fig. 1.9 shows a typical longitudinal acceleration $a(t)$ behavior acquired by accelerometer near the center of gravity during a full frontal impact. The value of the maximum acceleration and the impact duration depends upon the initial impact speed V.

The acceleration curve is generally obtained filtering the accelerometer data. The cutoff frequencies of the filter and the procedures are defined by SAEJ211 standard. Four channel frequency classes (CFCs) are defined, with different cutoff frequency. For total vehicle comparison, collision simulation input, and barrier face force, the cutoff frequency is 60 Hz, while integration for velocity or displacement is 180 Hz. A four-pole linear phase Butterworth filter is used, described by the following difference equation:

$$Y[t] = a_0X[t] + a_1X[t-1] + a_2X[t-2] + b_1Y[t-1] + b_2Y[t-2] \qquad (1.1)$$

where $X[t]$ is the input data sequence; $Y[t]$ is the filtered output data sequence; a_0, a_1, a_2, b_1, b_2 are the filter coefficient, dependent on CFC; and t is the sampling rate in second.

The filter coefficients are calculated following ISO 6487:

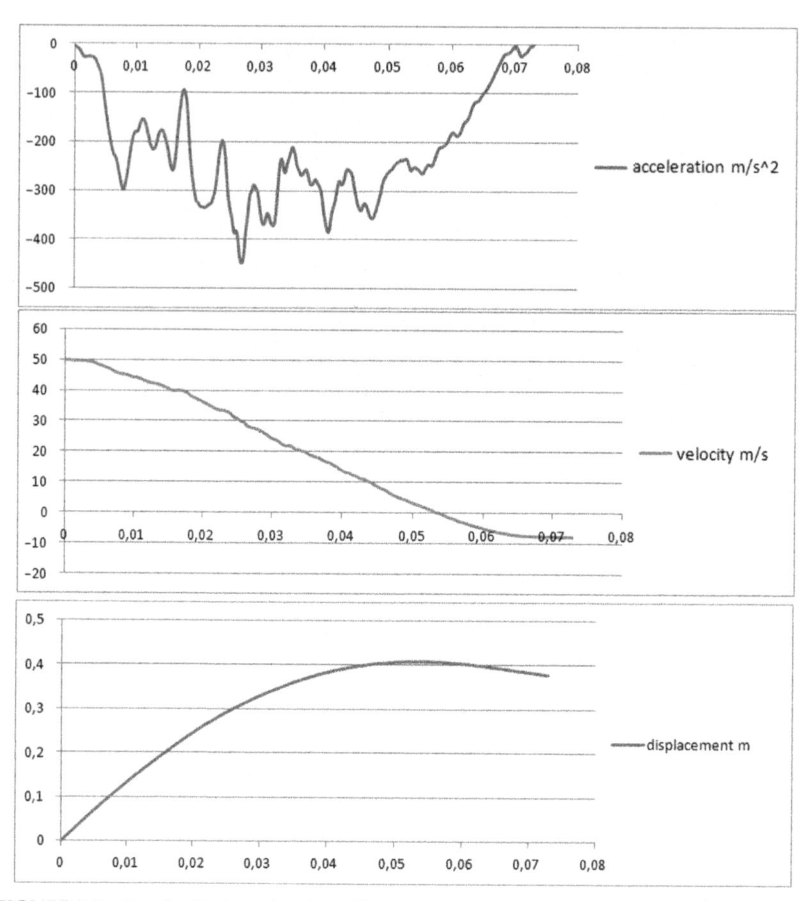

FIGURE 1.9 Longitudinal acceleration $a(t)$ versus time (s) measured during the impact against the barrier and velocity and space curves, obtained by numerical integration of the acceleration curve.

$$
\begin{aligned}
w_d &= 2\pi \mathrm{CFC} 2.0775 \\
w_a &= \frac{\sin(w_d T/2)}{\cos(w_d T/2} \\
a_0 &= \frac{w_a^2}{1+\sqrt{2}w_a + w_a^2} \\
a_1 &= 2a_0 \\
b_1 &= \frac{-2(w_a^2-1)}{1+\sqrt{2}w_a + w_a^2} \\
b_2 &= \frac{-1+\sqrt{2}w_a + w_a^2}{1+\sqrt{2}w_a + w_a^2}
\end{aligned}
\tag{1.2}
$$

The difference equation describes a two-pole filter. To achieve a four-pole filter, the data must pass through the two-pole filter twice: once forward and once backward, to prevent phase shifts.

Special techniques must be applied to eliminate the filter start-up effects from the filtered data set.

In Fig. 1.9 the speed $V(t)$ and space $x(t)$ curves over time are also reported, obtained by numerical integration of the acceleration curve acquired at the center of gravity location. In the case of a rigid barrier, space represents the vehicle center of gravity movement during the front deformation and, since the barrier is immovable, in case of negligible change of position of the center of gravity during the deformation, it coincides with the crushing of the vehicle itself. If the barrier is crushable, the space obtained by integration is the mutual crush, which is the sum of vehicle and barrier crush.

1.2.1 Centroid time

The space–time curve can also be obtained from the accelerometer data and only the first integral with the so-called moment-area method. This method yields a kinematic relationship between the maximum dynamic crush X, the corresponding velocity change, and the crash pulse centroid time t_C. The centroid C is the geometric center of the area defined by the acceleration curve from time zero up to the time of dynamic crush t_x.

Given a generic acceleration–time curve, assuming $V(0) = V_0$, by integration, we obtain the velocity–time curve, like the one shown in Fig. 1.10.

The displacement change $x_1 - x_0$, from t_0 until the time t_1, is the area under the $V(t)$ curve. The area can be computed as the rectangular area $V_0 t_1$ plus the area under the $t(v)$ curve:

$$x_1 - x_0 = V_0 t_1 + \int_{v_0}^{v_1} (t_1 - t) dV \tag{1.3}$$

as $dV = adt$, assuming differential area $adt = dA$, $t_1 - t = y$ we get

$$\int_{v_0}^{v_1} (t_1 - t) dV = \int_{t_0}^{t_1} (t_1 - t) adt = \int y dA \tag{1.4}$$

Eq. (1.4) represents the static moment with respect to the t_1 axis of the area under the curve $a(t)$ from 0 to t_1.

Let $t_1 - t^* = Y^*$, we can write

$$\int y dA = Y^* A = (t_1 - t^*) \int_0^{t_1} adt \tag{1.5}$$

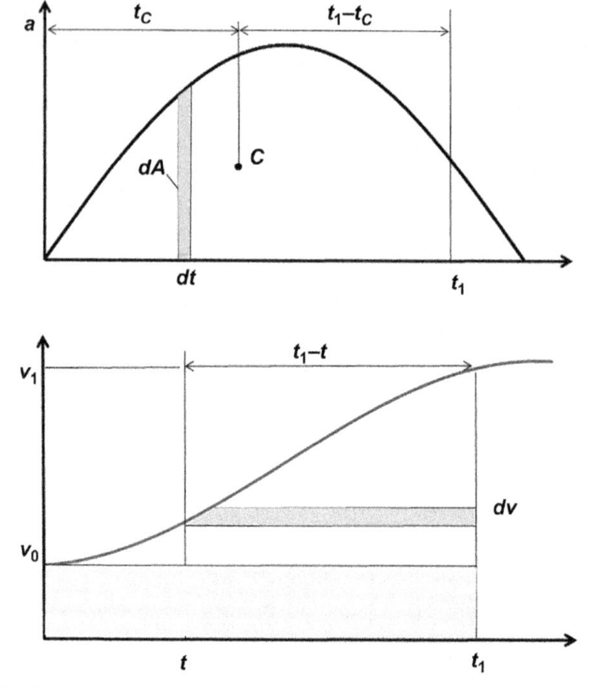

FIGURE 1.10 Graphical illustration of a moment-area method and displacement equation: $a(t)$ curve and the $V(t)$ curve obtained by integration.

t^* is the time corresponding to the geometric center of the area defined by the acceleration curve from time zero up to the time t_1. Replacing Eq. (1.5) by Eq. (1.3), we get

$$x_1 - x_0 = V_0 t_1 + (t_1 - t^*) \int_0^{t_1} a dt \tag{1.6}$$

where the integral is the area under the $a(t)$ curve between 0 and t_1.

Eq. (1.6) is the displacement equation, obtained without using the double integral of the accelerometer data.

The centroid time t_C is the time at the geometric center of the area of the crash pulse from time zero to the time of dynamic crush X and can be computed using Eq. (1.6).

In a fixed rigid barrier impact, at dynamic crush $v = 0$, and hence

$$\int_0^{t_1} a dt = V_1 - V_0 = 0 - V_0 = -V_0 \tag{1.7}$$

as $x_0 = 0$, from Eq. (1.6), we have

$$X = V_0 t_1 + (t_1 - t^*)(-V_0) = V_0 t^* = V_0 t_C \tag{1.8}$$

and the centroid time is simplified as

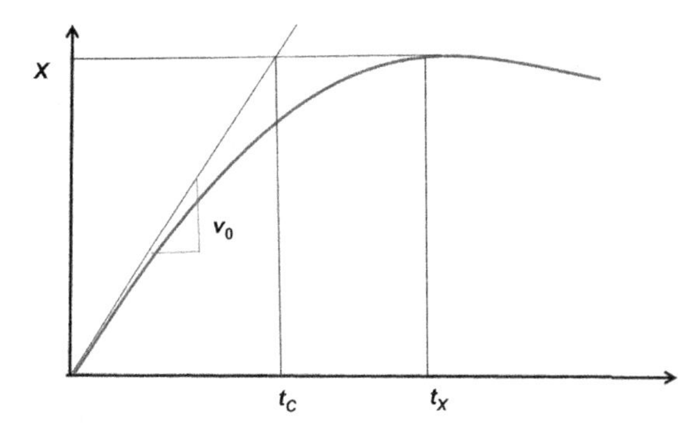

FIGURE 1.11 Graphical construction to compute the centroid time.

$$t_C = \frac{X}{V_0} \tag{1.9}$$

Since X is the dynamic crush and V_0 is the vehicle-rigid barrier impact velocity, t_C is the amount of crush per unit of impact speed and is defined as the characteristic length of the vehicle structure.

The reciprocal of t_C, namely b_1, is related to the vehicle stiffness, which is presented in Chapter 3, Models for the structural vehicle behavior.

The centroid time can be constructed for a given displacement-time history of a crash test as shown in Fig. 1.11, drawing a slope at time zero (the slope is the initial barrier impact velocity, V_0); the slope intersects the horizontal line through dynamic crush X at time t_C.

In a general case of a vehicle-to-vehicle impact, the centroid times of both vehicles are the same. In this case the centroid time can be computed using Eq. (1.9), where X represents the mutual crush (the sum of the dynamic crash of both vehicles) and V_0 is the closing speed of the two vehicles.

1.3 Force–deformation curve

The contact force with the barrier during the impact can be measured through load cells placed on the barrier itself.

In Fig. 1.12 the resultant of the forces measured by the load cells placed on the barrier is shown, for the impact referred to in Fig. 1.8.

However, under opportune simplifying assumptions, it is possible to evaluate the force/time curve starting from accelerometer data (Vangi, 2008; Huibers and De Beer, 2001). The assumptions are as follows:

- The mass of the vehicle remains constant during the impact; there are no detachments of parts with significant mass.
- During the collision, the moving mass is equal to that of the entire vehicle. Actually, as already noted, a part of the structures interested by the plastic flow (as the engine) already stops after the first contact with the barrier and the active mass reduces. This reduction is partly compensated by the mass of the occupants, coming into effect later in the crash.

- The vehicle deformations do not alter the position of the center of gravity of the vehicle significantly.
- In case of a full frontal impact, the acceleration is acquired near the center of gravity and, in case of eccentric impact, near the B pillar of the impacted side.

Under these assumptions, from the acceleration curve $a(t)$, it is possible to approximate the curve of the overall force acting on the vehicle, $F(t)$, by multiplying the acceleration values for the vehicle mass, following the first Newton's law:

$$F(t) = ma(t) \tag{1.10}$$

By eliminating the time variable from the $F(t)$ and $x(t)$ curves, the force as a function of the displacement $F(x)$ curve is obtained. As the barrier is fixed, the displacement coincides with the vehicle deformation plus, eventually, the deformation of the barrier.

The correlation between load cell data on the barrier and from the multiplication of mass times acceleration looks generally good. Fig. 1.13 shows,

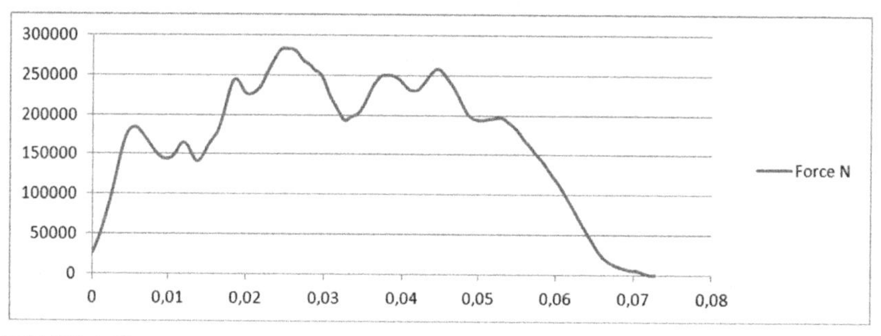

FIGURE 1.12 Resultant of the contact forces during a collision, as measured by load cells placed on the barrier.

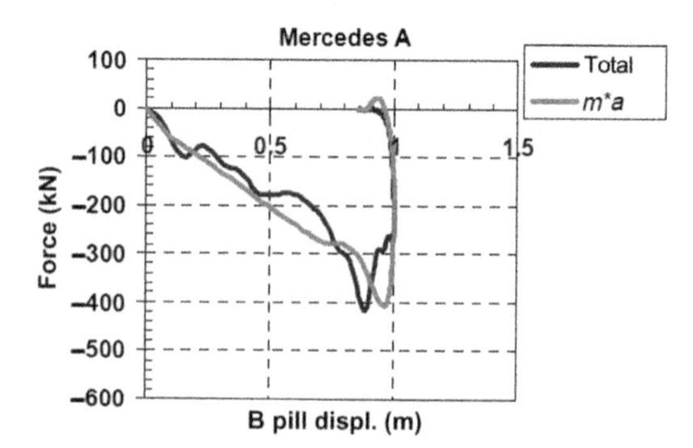

FIGURE 1.13 Comparison between force–deformation curves obtained from the load cell data and from the multiplication of mass times acceleration. *From Huibers, J., De Beer, E., 2001. Current front stiffness of European vehicles with regard to compatibility. In: Proceedings of the 17th ESV Conference, Paper No. 239.*

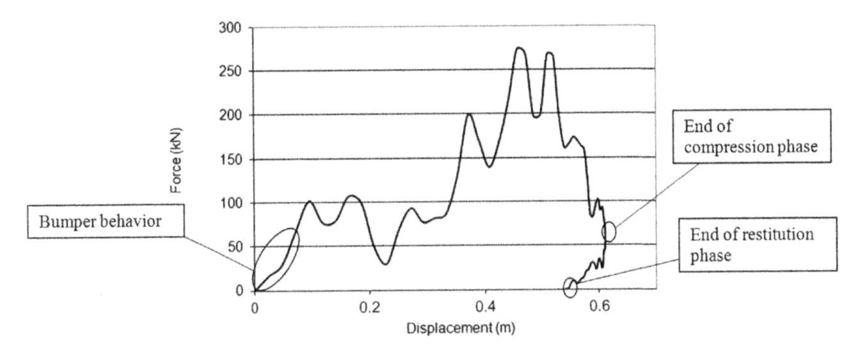

FIGURE 1.14 Force–deformation curve.

as an example, a comparison between the curves obtained from the load cell data and *ma* for a Mercedes class A, in a 40% offset impact against a deformable barrier (like EuroNCAP test).

Let us analyze a curve as in Fig. 1.14, for example, where a $|F(x)|$ curve is obtained from the accelerometer data.

Two distinct portions characterize this curve. The first phase is related to the compression of the structures, the deformation increases monotonically, although the force, due to the inhomogeneity of the structures involved, may have oscillatory behavior, with more or less broad peaks, and reached the maximum deformation. The second is the unloading phase or elastic restitution, in which the deformation decreases, with a partial recovery due to the elasticity of the structure. The $F(x)$ curve, in the phase of restitution, continues with a trend that depends, in general, on the maximum crush reached in the compression phase or, equivalently, by the impact speed.

The initial portion of the $F(x)$ curve is related, in front or rear impact, to the structural response of the bumper and the structures connected to this. In minor impact, up to 4 km/h of impact speed against the barrier, the bumper must absorb the energy without permanent damage, as stated by ECE R42 regulation. Then the slope of the first section of the $F(x)$ curve depends on the stiffness of the bumper. For higher forces, when the plastic yielding of the support bar occurs, the curve shows a more soft behavior.

The area under the $F(x)$ curve, from the beginning stage of the contact to the stage of maximum compression, represents the energy E_a globally absorbed during the impact, while the area under the curve related to the elastic restitution represents the E_r elastic potential energy recovered at this stage. The difference between these two energies indicates the overall energy E_d dissipated by the deformation of the vehicle.

The ratio between the areas subtended by the curve in the compression and restitution phases, under square root, is equal to the coefficient of restitution:

$$\varepsilon = \sqrt{\frac{E_r}{E_a}} \tag{1.11}$$

1.4 Distribution of the impact forces over time and space

The instrumented barriers allow measuring not only the resulting force but also the distribution of forces on the impact surface during the impact. Fig. 1.15 shows an instrumented barrier, composed of many load cells arranged on the impact surface.

Summing all the load cell values, the resultant force is obtained. The average height of force that a vehicle imparts on a barrier is shown in Fig. 1.16.

During the impact, the average height of force is generally changing, as illustrated in Fig. 1.17 for two types of vehicle, a family car and an SUV.

FIGURE 1.15 Exploded view of an instrumented barrier. From the left, impact block, assembly plate, fixed barrier, backing plate, load cells, and protective wooden plates. *From https://www.messring.de.*

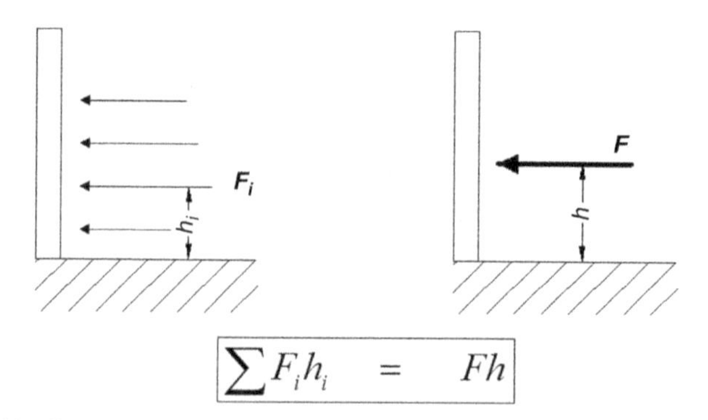

FIGURE 1.16 Calculation of the average height of force that a vehicle imparts on a barrier.

More generally, during the impact, the distribution of forces changes. In Figs. 1.18 and 1.19 the distribution of forces for an SUV at 34 and 56 ms from the beginning of the impact is shown. The peak of the force is in correspondence with the rails. In Fig. 1.20, on the other hand, the distributions of the impact forces are shown for two different vehicles, a family car and an SUV.

1.5 Parameters of influence for the force–deformation curves

The $F(x)$ curves describe the macroscopic structural response of each part of the vehicle (front, rear, etc.) in a given configuration of impact. The $F(x)$

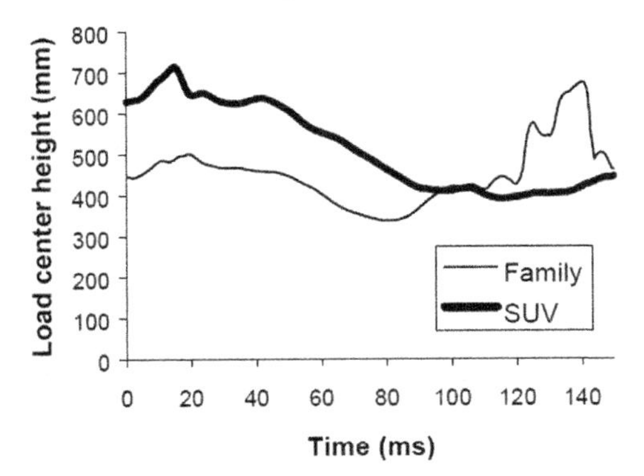

FIGURE 1.17 Variation of the center of force height showing a substantial difference between family car and SUV. *From Edwards, M., Happian-Smith, J., Davies, H., Byard, N., Hobbs, A., 2001. The essential requirements for compatible cars in frontal collisions. In: 17th International Technical Conference on the Enhanced Safety of Vehicles (ESV), Amsterdam, Holland (Edwards et al., 2001).*

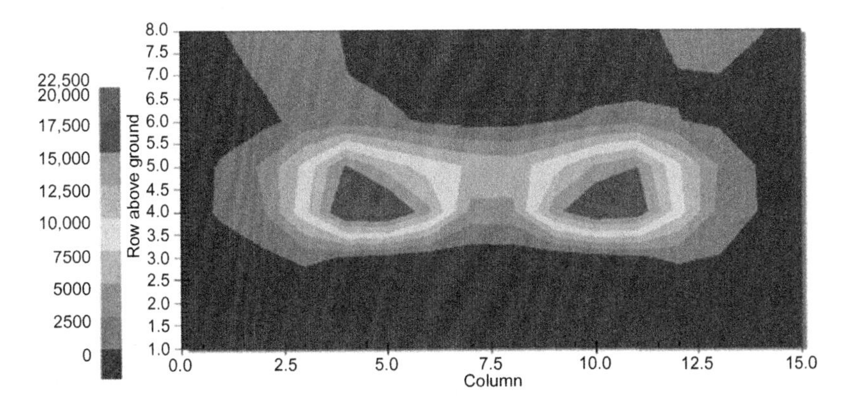

FIGURE 1.18 SUV force contour at 34 ms. *From Jerinsky, M.B., Hollowell, W.T., 2003. NHTSA's review of high-resolution load cell walls' role in designing for compatibility. In: Proceedings of the 18th International Technical Conference on the Enhanced Safety of Vehicles (ESV), Japan (Jerinsky and Hollowell, 2003).*

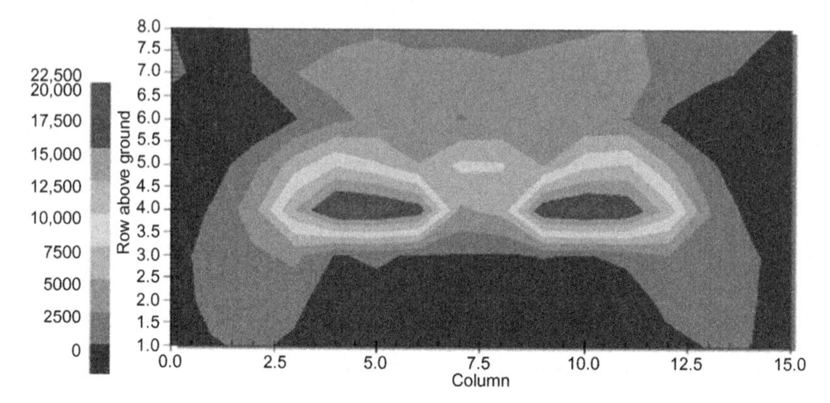

FIGURE 1.19 SUV force contour at 56 ms. *From Jerinsky, M.B., Hollowell, W.T., 2003. NHTSA's review of high-resolution load cell walls' role in designing for compatibility. In: Proceedings of the 18th International Technical Conference on the Enhanced Safety of Vehicles (ESV), Japan.*

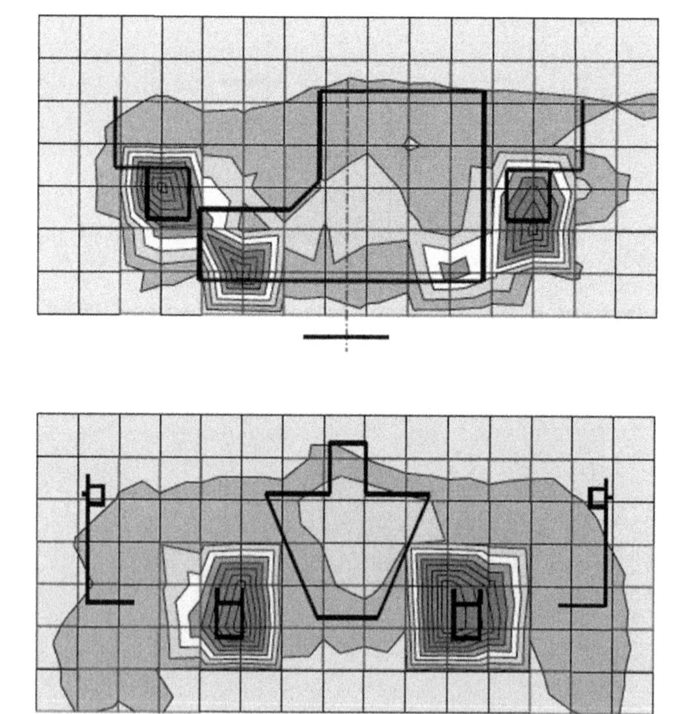

FIGURE 1.20 Comparison of load cell peak forces for a family car (above) and an SUV (below). Note, the main vehicle structure is superimposed on plots to show load paths. *From Edwards, M., Happian-Smith, J., Davies, H., Byard, N., Hobbs, A., 2001. The essential requirements for compatible cars in frontal collisions. In: 17th International Technical Conference on the Enhanced Safety of Vehicles (ESV), Amsterdam, Holland.*

curves, although generally obtained in tests against a rigid or deformable barrier, can be applied with good approximation even in collisions with other vehicles. That is, within certain limits, to have the same amount of deformation in a given configuration of impact, the same force must be applied, regardless of whether the object against the vehicle collides is a rigid barrier or another vehicle. This curve is, therefore, a characteristic of the vehicle. However, the $F(x)$ curves depend on the vehicle model, configuration of impact (with offset, angled, override/underride, etc.), and only marginally by the speed of impact.

1.5.1 Vehicle model

Different structural behaviors and, therefore, different $F(x)$ curves can be obtained depending on the vehicle model, on how it is built and with which architecture. Macroscopically, the vehicles can be grouped into classes, such as the medium-size family cars (like the Volkswagen Golf or the Toyota Corolla), the large saloon/executive cars (such as the Audi A6 or the Mercedes E200), MPVs (such as the Volkswagen Sharan or the Chrysler Voyager), and small family cars (such as Volkswagen Polo or the Peugeot 206).

For each category, the vehicle behavior is quite comparable, with some exceptions. By way of example, the results obtained by Huibers and De Beer (2001) are reported, analyzing crash tests with a 40% offset, with a deformable barrier. The force has been obtained from a load cell or multiplying the measured acceleration with the mass of the test vehicle. Trends for force displacements derived from acceleration times mass are the same as those derived from load cell data.

The displacement was obtained by double integration of acceleration and, therefore, also includes the crushing of the barrier. Fig. 1.21 shows, for example, the $F(x)$ curves for some models of large saloon/executive cars, while Fig. 1.22 shows the average $F(x)$ curves for each vehicle category. The load shows, for all the curves, an almost flat trend at the deformation between 0.2 and 0.4 m. This trend is associated with the barrier, which over a certain deformation behaves like a perfectly plastic material. After 0.4 m, the barrier reaches its maximum deformation, and the force rises rapidly due to the stiffness of the vehicle.

1.5.2 Offset

A collision that involves only a part of the front (collision with offset), for example, gives a different response from a collision in which the whole front is involved, as the resistance of the structures in each point is different.

The overlap percentage determines which parts of the frontal structure of the car are hit and contribute to the energy absorption, see Table 1.1. In case

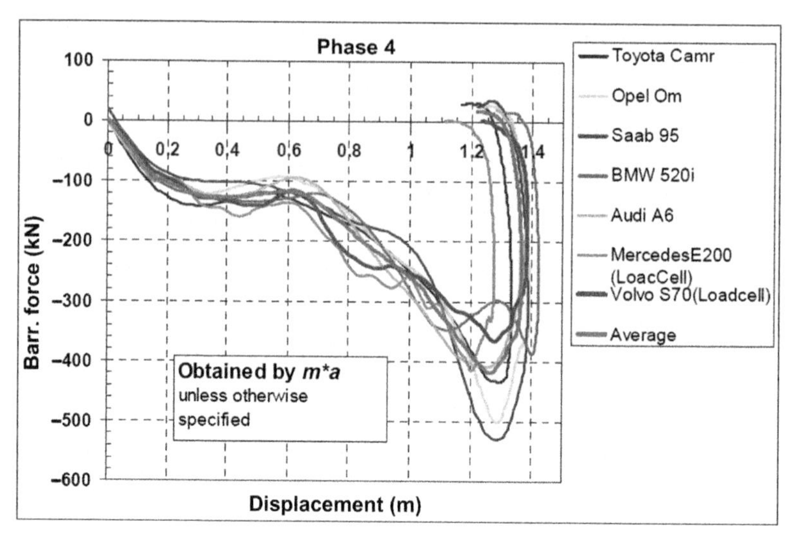

FIGURE 1.21 $F(x)$ curves for some large saloon cars. *From Huibers, J., De Beer, E., 2001. Current front stiffness of European vehicles with regard to compatibility. In: Proceedings of the 17th ESV Conference, Paper No. 239.*

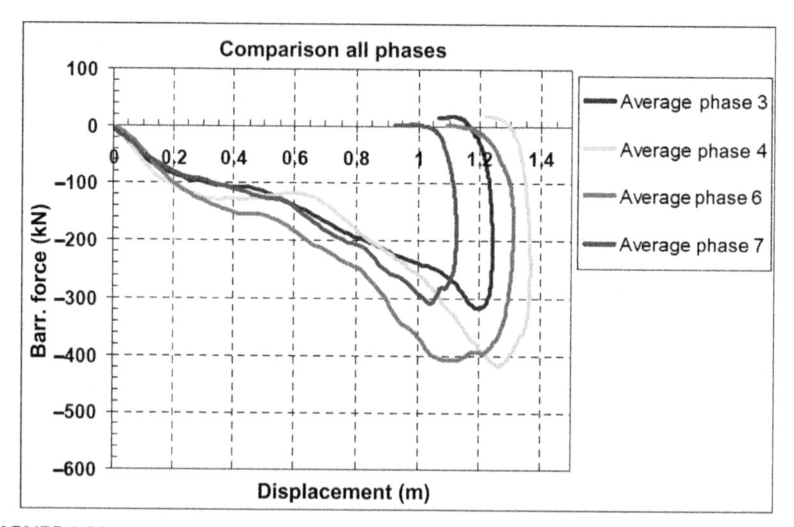

FIGURE 1.22 Average $F(x)$ curves for a different vehicle category (phases). *From Huibers, J., De Beer, E., 2001. Current front stiffness of European vehicles with regard to compatibility. In: Proceedings of the 17th ESV Conference, Paper No. 239.*

of a full overlap against a rigid wall, the two stiff longitudinal members and the motion of the engine can absorb most of the energy. During the first half of the crash duration, mainly the longitudinal members are loaded. In the second half, the engine is loaded as well.

A crash against a stiff pole can be regarded as a crash with a small overlap against a rigid wall. Only one stiff part, for example, one of the longitudinals or the engine, is hit.

The offset entity has a different influence on the deformations in the cases of low and high speeds. Figs. 1.23 and 1.24 show the numerical

TABLE 1.1 Different stiff parts involved in a different frontal overlap percentage.

Frontal overlap percentage	Stiff parts in the structure
70–100	2 Longitudinals + surrounding structure + engine/ firewall
40–70	1 Longitudinal + surrounding structure + engine/ firewall
30–40	1 Longitudinal + surrounding structure

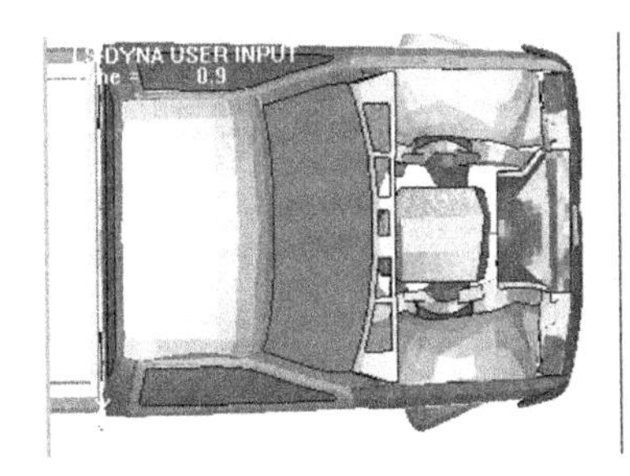

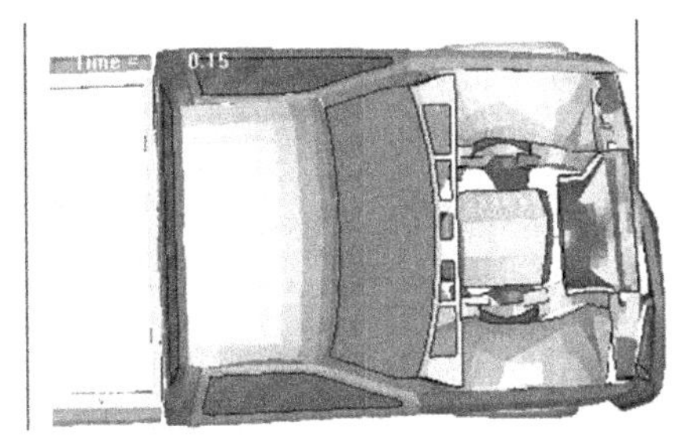

FIGURE 1.23 Numerical simulation of a low-speed collision (15 km/h), with 100% of the offset (left) and to 40% of the offset (right).

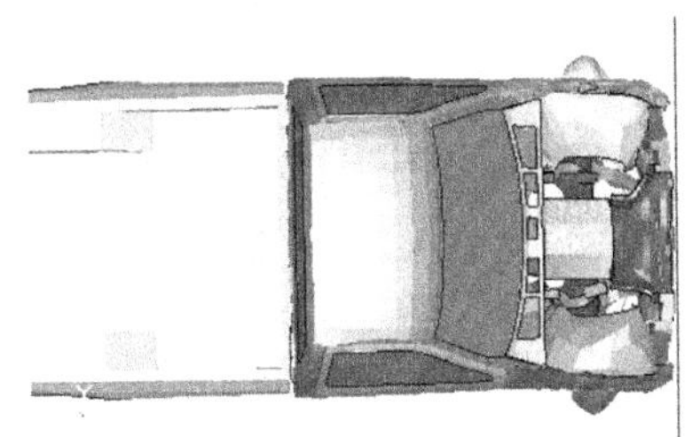
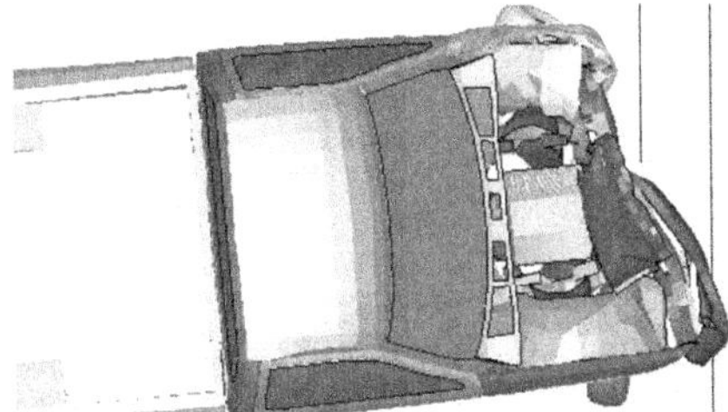

FIGURE 1.24 Numerical simulation of a high-speed collision (45 km/h), with 100% of the offset (left) and to 40% of the offset (right).

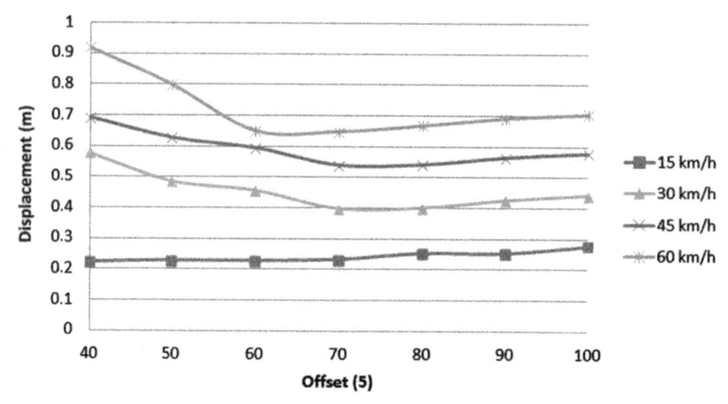

FIGURE 1.25 Magnitude of the maximum deformation varying the offset, for several impact speed against a rigid barrier.

simulations of low- and high-speed impacts, respectively, with 100% and 40% of offset. As can be seen, at low speeds, the structure predominantly interested by the plastic flow is that of the bumper while at a higher speed, other structures, such as spars, the engine, and the firewall, are also interested.

Fig. 1.25 shows the behavior of maximum deformation as a function of the offset, for several impact speeds, for the same vehicle of Figs. 1.23 and 1.24. It is observed that for low impact speed, the larger the offset, the more significant the deformation, since there it has high bending of the central part of the bumper. When the speed is such that the structures behind the bumper are involved, low offset impacts produce higher deformations, constituting a significant challenge for the crashworthiness of the vehicle.

Fig. 1.26 shows, for the same vehicle model and speed of impact against barrier of 15 km/h, $F(x)$ curves obtained at different levels of offset. It is noted that the curves differ from each other for their shapes, in particular in the maximum force values and the achieved deformations.

In the front or rear impact with 40% of overlap, there is substantial symmetry of the vehicle from a structural point of view, and the deformation affects nearly half of the front or rear. In this case, as a first approximation, one can assume that the force, for each value of the deformation, is equal to a comparable percentage of the force obtained in a full overlap collision, as

$$F_{100\%} = \frac{F_{40\%}}{0.4} \tag{1.12}$$

The greater the deformation (i.e., for high-speed impact), the better the approximation.

1.5.3 Oblique impact

Varying the angle with which the vehicle hits the barrier, as shown in Fig. 1.27, the direction of the forces also varies, and structures involved are stressed differently. Fig. 1.28 shows, for three vehicle models, the angle of the resultant of the impact forces (principal direction of force—PDOF) obtained from numerical simulation, varying the impact angle against the barrier. It is noted that the PDOF macroscopically has a linear trend and is always less than the inclination of the barrier since there are, in addition to the normal forces, also the frictional forces, parallel to the surface.

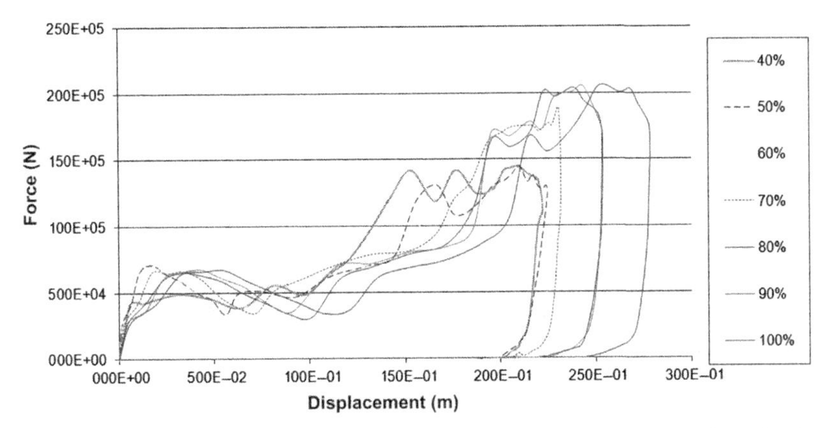

FIGURE 1.26 *F*(*x*) curves obtained for different offset values in an impact against a barrier, for the same model vehicle at a speed of 15 km/h.

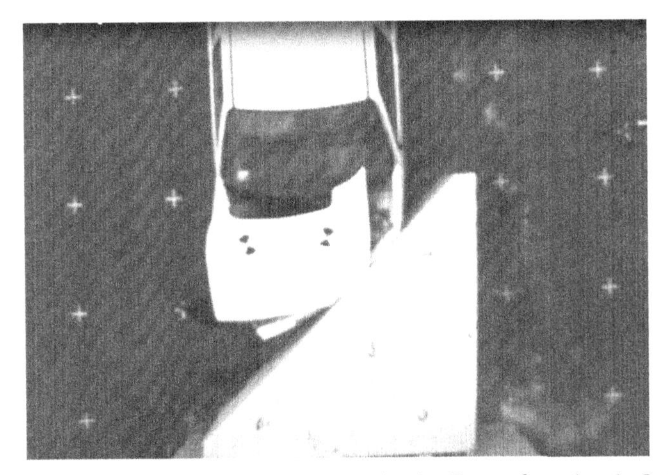

FIGURE 1.27 Crash test against 45-degree angled barrier. Test performed at the LaSIS (road safety and accident analysis laboratory)—University of Florence.

Collisions with different angles with respect to the barrier are characterized by different $F(x)$ curves, as shown in Fig. 1.29.

1.5.4 Underride/override

In different types of car impacts, as SUV-to-car or pickup-to-car, a strong misalignment of the bumper (underride or override) can occur. In such cases,

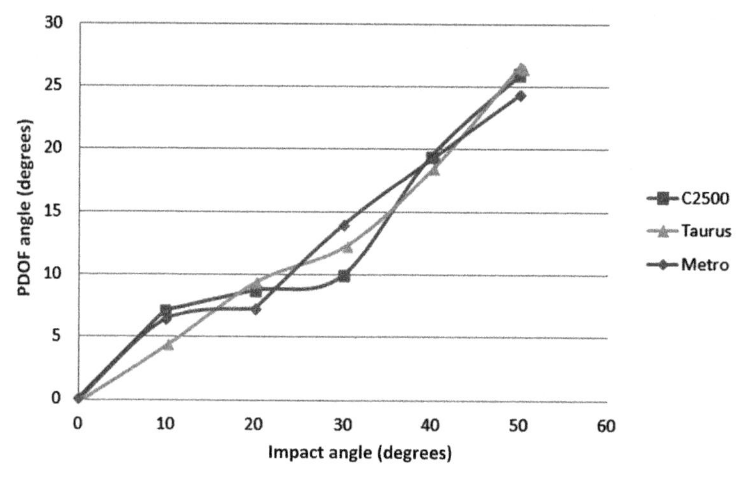

FIGURE 1.28 Behavior of the PDOF angle in the function of the angle of inclination of the barrier, for three different vehicle models. *PDOF*, Principle direction of force.

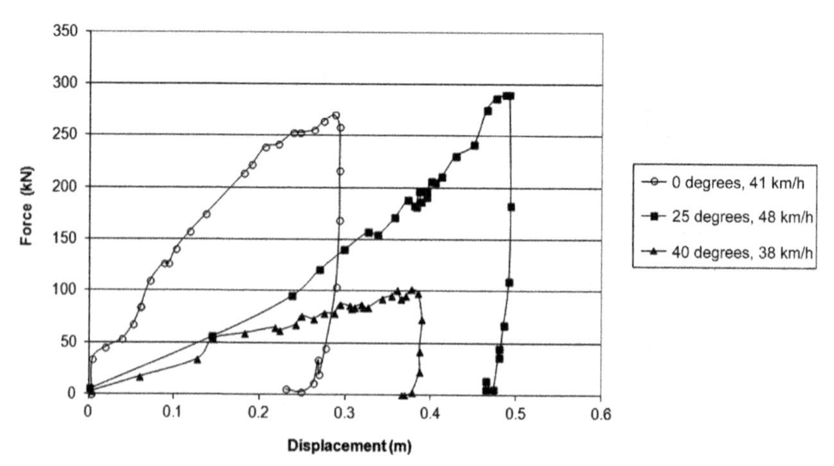

FIGURE 1.29 Experimental behavior of $F(x)$ curves for different angles and speeds and impact zones.

hundreds of millimeters relative bumpers' height is common and, as a consequence, the structures involved are only those above or below the bumper and not the entire front.

This occurs also in a rear-end impacts, when, for example, the vehicle is braking, see Fig. 1.30.

In this case, the damaged area and the acceleration curves are different for engagement or underride impact, as shown in Figs. 1.31 and 1.32.

Other underride situations are when a passenger vehicle collides with the rear end or side of trucks, trailers, or buses not equipped with effective guards. In these cases the vehicle continues to travel beneath the taller chassis of the larger vehicle, as depicted in Fig. 1.33.

In the case of underride/override the contact will have high losses of kinetic energy due to sliding friction of the sheets between them, even in case of low permanent deformations. This aspect is more evident in low-speed impact, in which the magnitude of the friction phenomena has a higher relative weight on the entire portion of kinetic energy loss.

Engagement

Alignement of bumbers beams

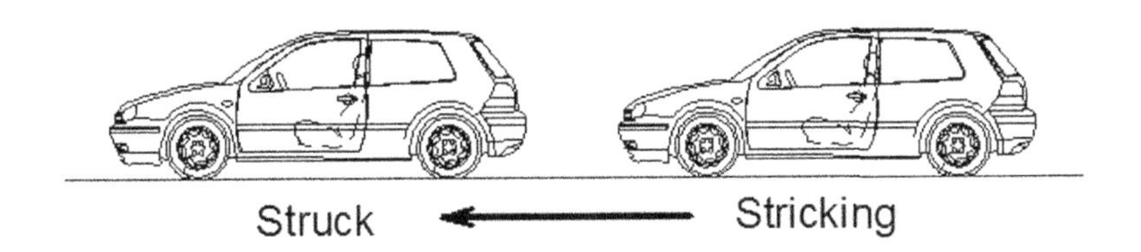

Underride

Misalignement of bumbers beams: stricking beam below struck beam

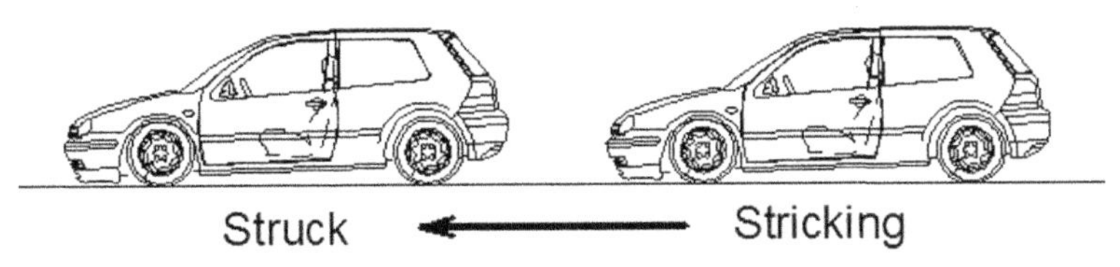

FIGURE 1.30 Different impact configurations in rear-end impacts depending on the braking of the vehicles.

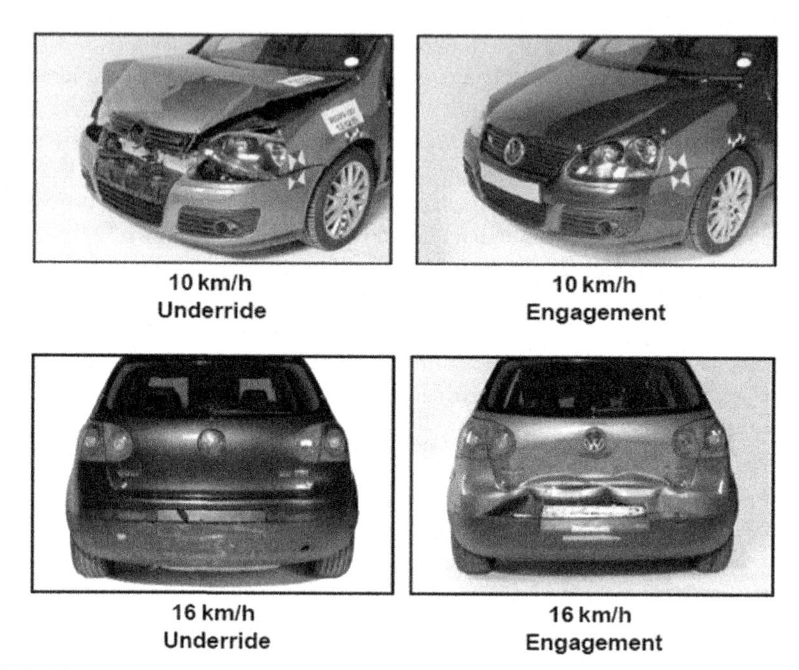

FIGURE 1.31 Different damage area involved in a striking and struck vehicle in the underride or engagement rear-end crash.

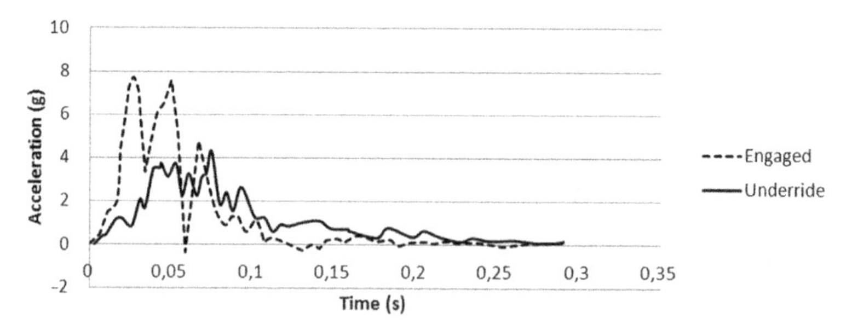

FIGURE 1.32 Target vehicle acceleration at 10 km/h impact speed: engagement versus underride.

In the case of underride/override the force/deformation curves are not comparable with the ones obtained in the standard crash test.

1.5.5 Impact speed

The $F(x)$ curve, in the typical range of speed that occurs in collisions between vehicles, is nearly independent of the deformation speed, that is,

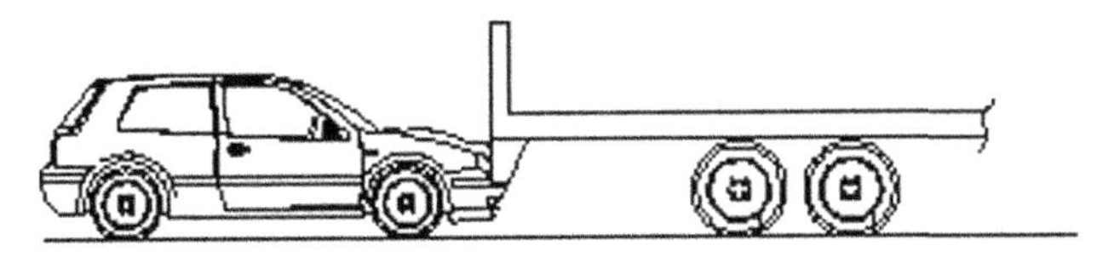

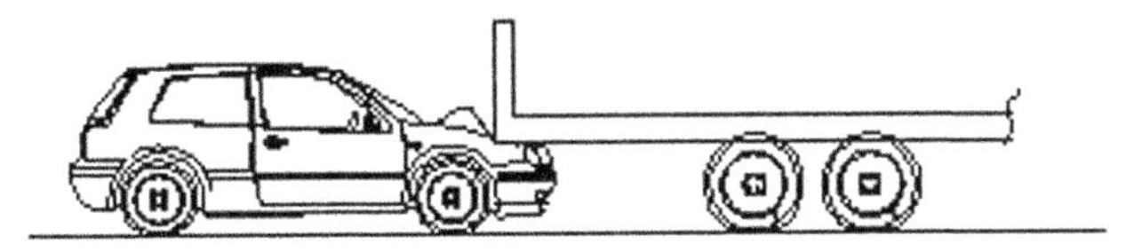

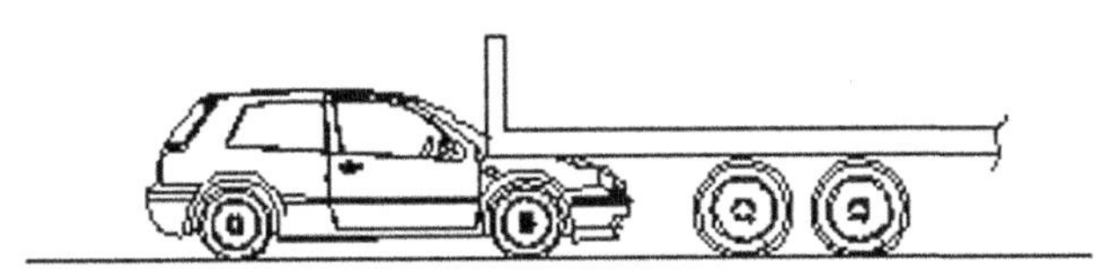

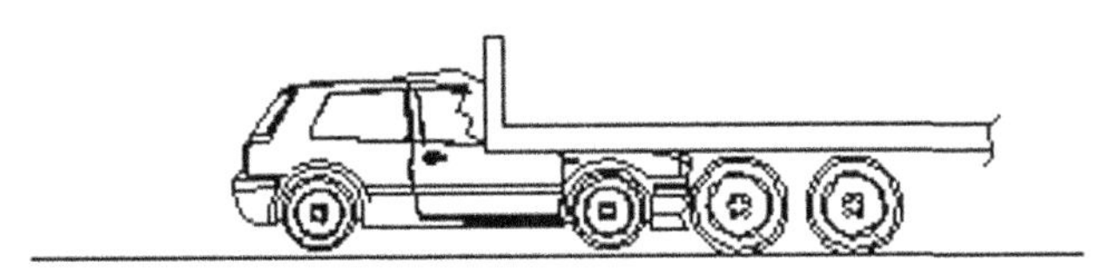

FIGURE 1.33 Different extent of underride in car-to-truck impact. *From Allen, K., 2010. The effectiveness of underride guards for heavy trailers. In: NHTSA Technical Report DOT HS 811 375 (Allen, 2010).*

from dx/dt. This means that during the phase of compression, increasing the deformation, the force follows, with a certain approximation, always the same trend. For example, starting from the vehicle acceleration curves obtained with different impact velocities, as shown in Fig. 1.34, $F(x)$ curves with the almost similar trend are obtained; obviously the higher the impact speed, the higher the maximum deformation reached. The differences found between the $F(x)$ curves obtained from various crash tests depend mostly on the following:

- Different vibrations recorded by accelerometers due to the elastic waves propagating inside the structure, the effect of which is not eliminated in the data preprocessing.
- The fact that the crash is driven by elastic instability phenomena and the structures buckling may not always show repeatable behaviors from one test to another due to small differences in impact configurations.

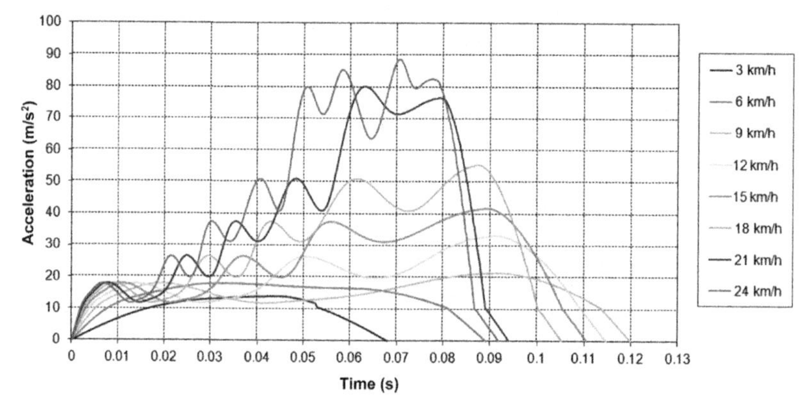

FIGURE 1.34 Acceleration curves obtained in head-on collisions at different initial impact speeds, for the same vehicle and impact configuration.

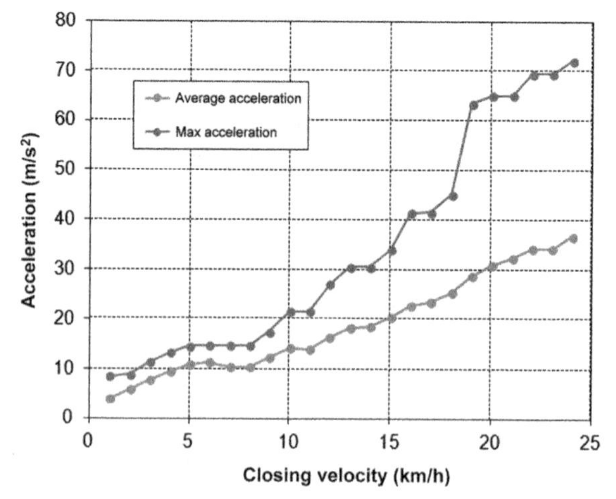

FIGURE 1.35 Peak (monotonic) and average (nonmonotonic) acceleration behavior increasing the impact speeds against barrier; results obtained from numerical simulation for a defined vehicle model.

At the macroscopic level, that is, at overall vehicle response under the impact, average acceleration may have a nonmonotonic behavior increasing the impact speed, as illustrated in Fig. 1.35. Exceeding certain threshold values, some structures are subject to yielding, with a lengthening of the impact time, lowering the average acceleration.

References

Allen K., 2010. The effectiveness of underride guards for heavy trailers. In: NHTSA Technical Report DOT HS 811 375.

Brantman, R., 1991. Achievable optimum crash pulses for compartment sensing and airbag performance. In: 13th International Technical Conference on Experimental Safety Vehicles (ESV), Paper S9-O-22, Paris, France, pp. 1134–1138.

Edwards, M., Happian-Smith, J., Davies, H., Byard, N., Hobbs, A., 2001. The essential requirements for compatible cars in frontal collisions. In: 17th International Technical Conference on the Enhanced Safety of Vehicles (ESV), Amsterdam, Holland.

Fenton, J., 1999. Handbook of Vehicle Design Analysis. SAE International, Warrendale, PA.

Griškevicius, P., Žiliukas, A., 2003. The absorption of the vehicles front structures. Transport 18 (2), 97–101.

Heisler, H., 2002. Advanced Vehicle Technology. SAE International, Warrendale, PA.

Huibers, J., De Beer, E., 2001. Current front stiffness of European vehicles with regard to compatibility. In: Proceedings of the 17th ESV Conference, Paper No. 239.

Jerinsky, M.B., Hollowell, W.T., 2003. NHTSA's review of high-resolution load cell walls' role in designing for compatibility. In: Proceedings of the 18th International Technical Conference on the Enhanced Safety of Vehicles (ESV), Japan.

Jones, N., 2012. Structural Impact. Cambridge University Press.

Stronge, W.J., 2000. Impact Mechanics. University of Cambridge, Cambridge University Press.

Vangi, D., 2008. Accident Reconstruction (in Italian). Firenze University Press, ISBN 978-88-8453-783-6.

Witteman, W.J., 1999. Improved vehicle crashworthiness design by control of the energy absorption for different collision situations. Technische Universiteit Eindhoven, Eindhoven. doi: https://doi.org/10.6100/IR518429.

Chapter 2

Impact impulsive models

Chapter Outline

2.1 Impact models

The impact of a vehicle against another vehicle or an obstacle has a typical duration of the order of 100 ms. This impact duration is large compared to the "shock-wave time." During the collision, in addition to the forces exchanged with the ground, the vehicle undergoes the impact forces, generally very significant. The magnitude and the trend over time of these forces depend on the vehicle and obstacle stiffness in addition to the relative speed and the configuration of impact. Due to these forces, the vehicle undergoes deformations, accelerations, and displacements, while the occupants are being solicited in several ways and interact with the restraint systems. The vehicle crash behavior can be described through the relationship between structures' stiffness, forces, and displacements, by applying the classical differential equations of motion. Such description can be made by considering the time as a continuous variable with the use of appropriate analytical force/deformation models of the vehicle, see for example, the lumped

Vehicle Collision Dynamics. DOI: https://doi.org/10.1016/B978-0-12-812750-6.00002-0

parameter models (Jonsén et al., 2009; Pahlavani and Marzbanrad, 2015; Pawlus et al., 2011b; Ofochebe et al., 2015; Marzbanrad and Pahlavani, 2011) or hybrid models (Pawlus et al., 2011a) or pulse model crashes (Iraeus and Lindquist, 2015; Varat and Husher, 2003).

The crash behavior can be described considering a discrete time domain, as occurs in multibody models (El Kady et al., 2016; Amato et al., 2013; Sousa et al., 2008) and finite-element models (Yildiz and Solanki, 2012; Hamza and Saitou, 2003; Chen et al., 2015). These approaches allow obtaining information on vehicle behavior in terms of deformation, displacements, speed, and accelerations, during the entire impact duration.

Another way to analyze the impact between two bodies is based on the use of impulsive models, that is, based on the use of the integrals of motion, momentum, and energy, in which only the initial and final instants of the impact are considered (Goldsmith, 2001; Kudlich, 1966; Vangi, 2008; Brach and Brach, 2005; Gilardi and Sharf, 2002; Brach, 1983; Ishlkawa, 1993; Huang, 2002; Han, 2015; Klausen et al., 2014).

The use of integrals of motion allows obtaining kinematic parameters with algebraic equations instead of differential equations. However, in these equations, time and position variables do not explicitly appear but only the conditions (typically velocity) at the beginning and end of the crash. Information about what occurs during the impact, as the way the forces during the impact are exchanged among the vehicles and its deformation is lost.

In the impulsive models, it is assumed that the impact forces are exchanged instantaneously and that the vehicle configuration is fixed; hence, the deformations and displacements of the vehicles during the impact can be neglected. Therefore the configuration assumed for the analysis of impact is the one at the maximum crush of the vehicles, which is generally corresponding to the maximum forces exchanged instantly.

In summary, the following assumptions are made in impulsive models:

- The duration of the impact is not considered.
- The system formed by the two vehicles is isolated, that is, all external forces, such as friction wheels—terrain, aerodynamic, and gravitational forces are small and negligible comparing to the impact forces.
- Conservation of kinematics configuration is assumed during impact.
- The deformation of the bodies during the impact is not counted in the calculation.

In impulsive models, vehicles can be schematized considering only some components of the motion or some forces involved in. In many practical cases, sufficiently approximate information can be obtained by considering the vehicles as material points, that is, models with one or two degrees of freedom, and by neglecting the external forces. In other cases, for example, in low-speed impacts, the external forces (Vangi and Mastandrea, 2005), such as those are obtained from the wheel–terrain contact should also be considered. If the angular speeds should be also evaluated, three-degrees-of-freedom models must be used (Vangi, 2008).

In this chapter, neglecting the aerodynamic forces and those exchanged with the terrain, an impulsive model with three degrees of freedom is described for vehicle impacts. The model is general, and the models in one or two degrees of freedom can be derived by taking the appropriate simplifications.

2.2 Contact plane, center of impact

In the collision between two vehicles, there is a contact surface, identifiable with the area where there has been direct contact between vehicles (the direct damage area of vehicles), through which the bodies exchange the impact forces. Generally, this contact surface is not flat, but for further analysis purposes, can be approximated as a plane, said *contact plane*. The position of the contact plane varies during the impact, as well as the forces vary in the direction and magnitude (Ishikawa, 1994).

In every instant the contact forces are equivalent to their resultant applied in a central axis point,[1] plus a moment lying in a plane orthogonal to it. Since the contact forces are oriented mainly parallel to the road, the magnitude of this moment, which tends to rotate the vehicle about the central axis, can generally be neglected. The contact forces, at each instant, are therefore equivalent to a single resulting force applied at a given point, as shown in Fig. 2.1.

During the collision, as the resultant force varies in both direction and magnitude, the central axis also varies, and then the point at which the resultant is instantly applied is varying as well. In Fig. 2.2, for two impact configurations between vehicles, the time variation of the point at which the resultant of forces is applied is shown.

For the analysis of the impact phase with an impulsive model, as we consider only the start and end instants, a point called *center of impact* can be considered. In this center of impact, one can think the average of the resultants forces is applied during the whole impact time, such as to produce the same effects on the vehicle's motion that the one produced by all the forces, variable in time and space.

The integral over time of the resultant of the contact forces is equal to the impulse **I**, and its direction is indicated as the principal direction of force (PDOF) in the reference of local coordinates (see Fig. 2.3).

---

1. The central axis, for a given system of forces, is a line parallel to the resultant of the forces. In every point of this line the resulting moment of the forces is reduced to the only part parallel to the resultant itself. In the case of forces lying all on one plane, the resulting moment with respect to the points of the central axis is null.

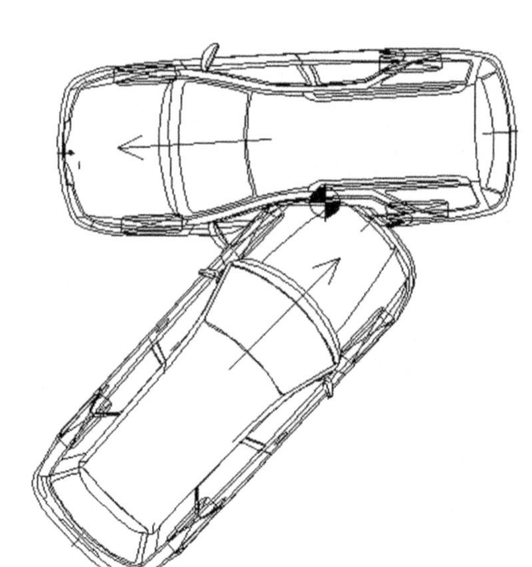

FIGURE 2.1 Point where the resultant force is applied at a given instant during the impact.

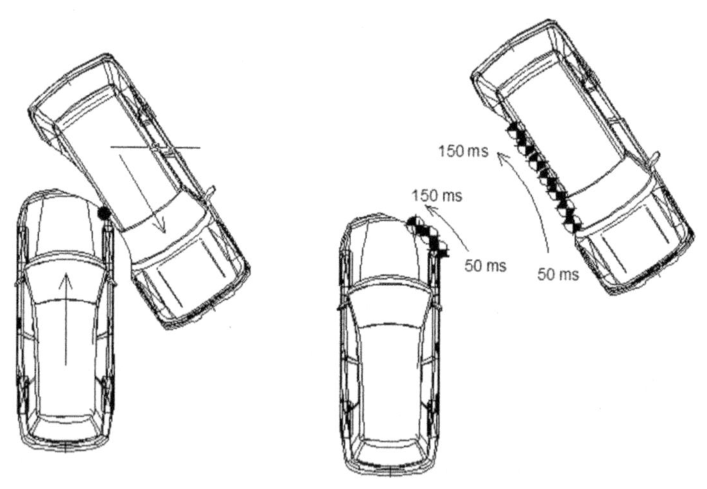

FIGURE 2.2 Example of variation of the point of application of the forces resultant during the impact for a given configuration between vehicles.

The center of impact can be approximately placed on the final deformed surface of the vehicles, in a central position of the area of direct damage. Some researchers (Kolk et al., 2016) express as a reasonable estimate of the position of the center of impact, the center of gravity of the area resulting from the intersection of the silhouettes of vehicles at the time of the maximum superposition.

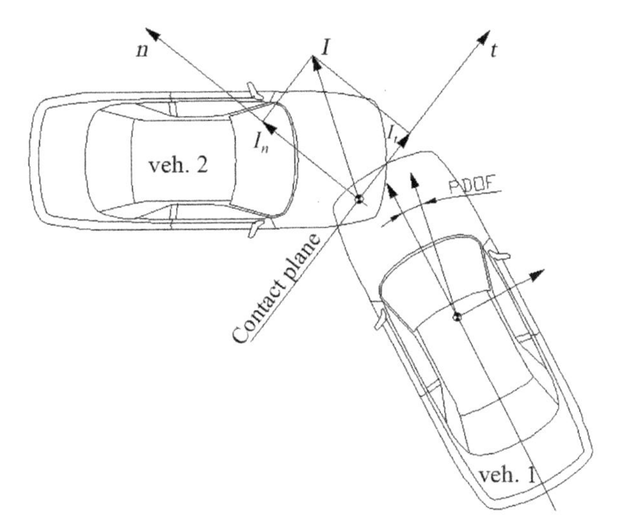

FIGURE 2.3 Contact plan, center of impact, resulting forces, and impulse components in the normal and tangential directions and PDOF referring to vehicle 1. *PDOF*, Principal direction of force.

2.3 Momentum, impulse, and friction coefficient

The impact forces and the impulses **I** acting on the vehicles are equal and of opposite directions, according to Newton's third law. In an isolated system the impulse matches the variation of momentum experienced by vehicles (impulse theorem):

$$\mathbf{I} = m_1 \mathbf{\Delta V}_1 = -m_2 \mathbf{\Delta V}_2 \tag{2.1}$$

Eq. (2.1) indicates that the change of momentum of the vehicle 1 is equal to that of the vehicle 2 but opposite in sign. Moreover, since Eq. (2.1) is a vector equation, it indicates that the velocity changes of the vehicle take place in the same direction of the impulse.

Adjusting the terms of Eq. (2.1), between the modulus of the vector, we can also write

$$\frac{\Delta V_1}{\Delta V_2} = -\frac{m_2}{m_1} \tag{2.2}$$

The relation shown in Eq. (2.2) is valid for every type of collision and indicates that the speed variations experienced by vehicles during the impact are inversely proportional to their respective masses, which means that a more massive vehicle undergoes a smaller variation of speed.

Making explicit and changing the order of the terms of Eq. (2.1), it is possible to express the conservation of momentum in the collision:

$$m_1 \mathbf{V_1} + m_2 \mathbf{V}_2 = m_1 \overline{\mathbf{V}}_1 + m_2 \overline{\mathbf{V}}_2 \tag{2.3}$$

where $\mathbf{V_1}$ and $\mathbf{V_2}$ indicate the speeds of the vehicles 1 and 2 at the beginning of contact, and $\overline{\mathbf{V}}_1$ and $\overline{\mathbf{V}}_2$ those at the end, that is, at the moment in which the two vehicles are separated. Eq. (2.3) is valid in every instant, for example, the amount of momentum of the system remains constant at each instant t:

$$m\mathbf{V_1} + m\mathbf{V_2} = \mathrm{cos}t \quad \forall t. \tag{2.4}$$

From Eq. (2.1), with some passages, we get

$$\mathbf{I} = m_c(\Delta\mathbf{V}_1 - \Delta\mathbf{V}_2) \tag{2.5}$$

where m_c is the "common mass," defined as

$$m_c = \frac{m_1 m_2}{m_1 + m_2} \tag{2.6}$$

Considering the meaning of relative velocity $\mathbf{V_R} = \mathbf{V_2} - \mathbf{V_1}$, the difference between the final and initial relative velocities is equal to the difference between the velocities variations $\mathbf{\Delta V}$ experienced by individual vehicles during impact:

$$\overline{\mathbf{V}}_\mathbf{R} - \mathbf{V_R} = \Delta\mathbf{V}_2 - \Delta\mathbf{V}_1 \tag{2.7}$$

Eq. (2.5) can be written as

$$\mathbf{I} = m_c(\mathbf{V_R} - \overline{\mathbf{V}}_R) \tag{2.8}$$

Eq. (2.8) allows identifying, with a good approximation, the impulse direction. In fact, since the relative velocity at the end of impact is generally small in comparison to the initial one (see also Section 2.3) the direction of $\mathbf{I}$ is practically coincident with the direction of the $\mathbf{V_R}$ (Marquard, 1962).

The impulse of the contact forces has a component I_n perpendicular to the contact plane and a component I_t parallel to it (see Fig. 2.3). The ratio between the normal and tangential components is

$$\mu = \frac{I_t}{I_n} \tag{2.9}$$

In some applications, μ may correspond directly to a coefficient of dynamic sliding friction, that is, Coulomb friction. However, it should be noted that the abovementioned definition is not subject to any limitations and permits modeling of such diverse processes as combinations of dry friction, inelastic shear deformation of materials, etc.

The coefficient μ can be positive or negative, depending on the sign of I_t. For the determination of its sign, it is sufficient to recall that the pulse I_t always tends to oppose to the relative motion of the two bodies.

Eliminating $\mathbf{\Delta V_2}$ between Eqs. (2.1) and (2.7) and acting on the components of the vectors along n and t, we obtain

$$
\begin{aligned}
\Delta V_{1n} &= \frac{m_2}{m_1 + m_2}\left(1 - \frac{\overline{V}_{Rn}}{V_{Rn}}\right)V_{Rn} \\
\Delta V_{1t} &= \mu\frac{m_2}{m_1 + m_2}\left(1 - \frac{\overline{V}_{Rn}}{V_{Rn}}\right)V_{Rn}
\end{aligned}
\tag{2.10}
$$

2.4 Coefficient of restitution

Let us consider a frontal impact between a vehicle 1 and a vehicle 2; the collision is characterized by a first compression phase in which the centers of gravity of the vehicles approach each other, with positive closing speed. This phase continues until a minimum distance is reached, that is, the maximum deformation, also called dynamic deformation X. At this instant, denoted by t_x, the speed of vehicles is equal, that is, the closing speed is zero. After that, there is the restitution phase, in which the closing speed is reversed in sign, and the vehicles move apart, elastically recovering part of the deformations.

The speed of the vehicles at the time t_x, being null the closing speed $\mathbf{V_R}$, coincides with the speed $\mathbf{V_G}$ of the center of gravity of the system:

$$
\mathbf{V_G} = \frac{m_1\mathbf{V_1} + m_2\mathbf{V_2}}{m_1 + m_2} \tag{2.11}
$$

As the system is isolated, the total momentum is constant over time, and the center of gravity speed can also be written in terms of the final speed:

$$
\mathbf{V_G} = \frac{m_1\overline{\mathbf{V}}_1 + m_2\overline{\mathbf{V}}_2}{m_1 + m_2} \tag{2.12}
$$

At the end of the contact between the vehicles, there is a permanent deformation, said residual crush.

In both compression and restitution phases the contact forces between the vehicles tend to repulse the vehicles and to reverse the initial sign of the closing speed.

Considering the intervals $t_{iniz} - t_x$ and $t_x - t_{\text{end}}$, the impulse theorem can be written as

$$
\begin{aligned}
\mathbf{I_c} &= m_1(\mathbf{V_1} - \mathbf{V_G}) = m_2(\mathbf{V_G} - \mathbf{V_2}) \\
\mathbf{I_r} &= m_1(\mathbf{V_G} - \overline{\mathbf{V}}_1) = m_2(\overline{\mathbf{V}}_2 - \mathbf{V_G})
\end{aligned}
\tag{2.13}
$$

Newton first postulated, based on experimental evidence, that the components along n of restitution and compression pulses was proportional and called coefficient of restitution their ratio, which is denoted by ε as follows:

$$\varepsilon = \frac{I_{rn}}{I_{cn}} \tag{2.14}$$

In the case of totally elastic collision, that is when there are no permanent deformations, $\varepsilon = 1$ and for each vehicle the variation of the velocity in the compression phase is equal and opposite to that in the restitution phase; the vehicles at the end of collision, therefore, possess the same velocity as at the initial time but changed in sign.

Conversely, if the collision is plastic, the dynamic deformation coincides with the residual one, without any elastic recovery, $\varepsilon = 0$, and the two vehicles at the end of the collision possess the same common velocity $\mathbf{V_G}$. The coefficient of restitution, being defined as the ratio between the pulse magnitude in the restitution phase and that in the compression phase, expresses the structural behavior of the two vehicles.

Substituting Eq. (2.14) into Eq. (2.13) we get

$$\varepsilon = \frac{I_{rn}}{I_{cn}} = \frac{(V_{Gn} - \overline{V}_{1n})}{(V_{1n} - V_{Gn})} \tag{2.15}$$

from which, substituting $\mathbf{V_G}$ with expressions given by Eqs. (2.11) and (2.12), we obtain

$$\varepsilon = \frac{\left((m_1\overline{V}_{1n} + m_2\overline{V}_{2n})/(m_1 + m_2)\right) - \overline{V}_{1n}}{V_{1n} - \left((m_1 V_{1n} + m_2 V_{2n})/(m_1 + m_2)\right)} \tag{2.16}$$

from which, simplifying, we obtain the most widely used form of the coefficient of restitution, indicated first by Poisson:

$$\varepsilon = -\frac{\overline{V}_{Rn}}{V_{Rn}} = \frac{\overline{V}_{2n} - \overline{V}_{1n}}{V_{1n} - V_{2n}} \tag{2.17}$$

The coefficient of restitution in the collision between two vehicles can be determined experimentally, based on crash tests. Typically, it assumes values close to zero for high-speed impact, characterized by large deformations, while it assumes higher values decreasing the closing speed. Experimental measurements of head-on collisions between vehicles (Antonetti, 1998) show, for impact with 5 km/h $< V_R <$ 15 km/h value of the coefficient of restitution in the range $0.2 < \varepsilon < 0.6$, while it can reach 0.8 for impact at very low relative speed, lower than 5 km/h. In practice the unit value is never reached due to viscous behavior of the material of the bumper and the presence of other dissipative external forces. In Fig. 2.4 the coefficient of restitution qualitative trend for frontal collisions between vehicles is shown, as a function of the initial closing speeds.

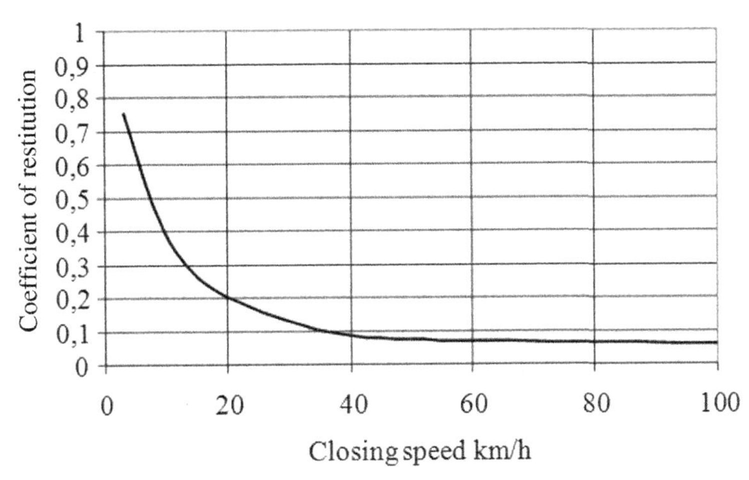

FIGURE 2.4 Qualitative trend of the coefficient of restitution in frontal impacts between two vehicles, in function of the initial closing speed.

Considering the definition of the coefficient of restitution (2.17), Eq. (2.10) becomes

$$\begin{aligned} \Delta V_{1n} &= \frac{m_2}{m_1 + m_2}(1+\varepsilon)V_{Rn} \\ \Delta V_{1t} &= \mu \frac{m_2}{m_1 + m_2}(1+\varepsilon)V_{Rn} \end{aligned} \tag{2.18}$$

Eq. (2.5), in the direction n and with the definition of the coefficient of restitution (2.17), becomes

$$I_n = m_c(1+\varepsilon)V_{Rn} \tag{2.19}$$

In the following, the three factors affecting the impact are grouped as

- closing speed, summarized in the term V_{Rn};
- vehicles masses, summarized in the term m_c; *and*
- structural aspects, summarized in the term ε.

These three factors indicate that the impulse intensity, or the forces between the two vehicles, depends only on the closing speed, on the common mass, and the stiffness of the vehicles.

2.4.1 Closing speed

The impact intensity depends directly on the closing speed. A collision that occurs between a stationary vehicle hit by a second vehicle at a speed of 20 km/h is equivalent, in terms of impulse and ΔV, to a collision between a vehicle moving at 80 km/h struck by a second moving at 100 km/h. Obviously, in the second case the postimpact motions of the vehicles are very different, but the extent of damage sustained by the vehicles, which depends on internal contact forces, is essentially the same.

2.4.2 Mass effect

The impact severity is influenced by a numeric combination of vehicle masses, expressed in the m_c term, said equivalent mass of the system. In this term the contributions of the individual masses are not explicitly identified; however, special cases can be analyzed where such contributions are more explicit. In the case of vehicles of the same mass m, for example, we get $m_c = m/2$. If one of the two masses can be considered infinite, as in the case of a collision against a wall or a barrier, then we have $m_c = m$. It is observed that m_c varies between two extreme values: from the mid to the entire mass of the lighter vehicle.

2.4.3 Structural aspects

The coefficient of restitution summarizes the effects of the structural characteristics of the vehicles. It can be stated that the type of vehicle, and in particular the geometry and the materials of its structures, defines the type of impact response and the percentage of energy restituted, compared to that absorbed.

2.5 Centered and oblique impacts

In the technical literature, there are several classifications for impacts, depending on if the arrival velocity of the vehicles is aligned with each other, or directed toward the center of gravity of the opposite vehicle, etc.

In the following a classification of impacts according to the impulse direction with respect to the centers of gravity of the vehicles is used.

Centered impact: the line of action of the impulse passes through the centers of gravity of both vehicles; in Fig. 2.5, two cases of centered impacts are shown. It is noted that the initial motion direction of the vehicles does not come under the classification.

Since the resultant of the forces is passing through the centers of gravity of the vehicles, there are no variations of the angular speed of the vehicles but only variation of the velocity. In this case, it is possible to use a two-degrees-of-freedom model of the vehicles, that is, a model in which the mass of the vehicle is considered concentrated in the center

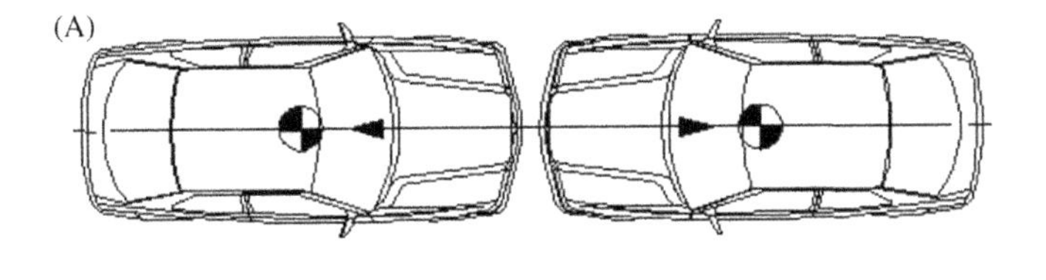

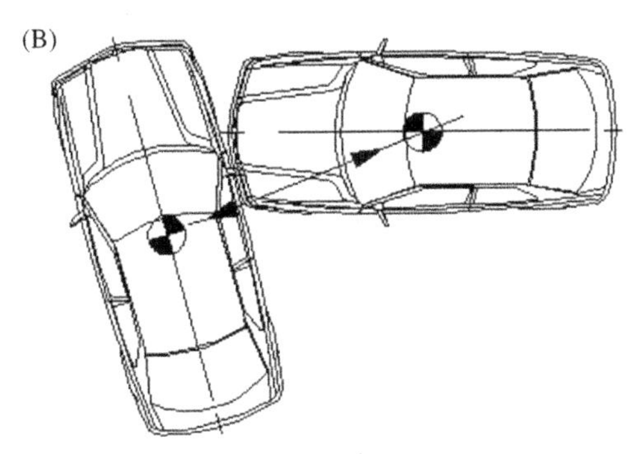

FIGURE 2.5 Centered impacts, in which the impulse line of action passes through the centers of gravity of both vehicles. (A) coaxial centered impacts (or direct); (B) oblique centered impacts.

of gravity and considering only the x and y coordinates of its plane motion.

An impact is centered if and only if the line of action of the impulse coincides with the alignment of the center of impact with the centers of gravity of the vehicles (see Fig. 2.6).

Oblique impact: the impulse line of action does not pass through the centers of gravity of the vehicles, as shown in Fig. 2.7. The impulse generates variations of the angular velocity around the vertical axis of the vehicles. To assess the extent of these rotations is necessary to use a rigid body model of the vehicle, with three degrees of freedom, the two x and y coordinates of the center of gravity, and the rotation angle around the vertical axis.

2.6 Model with three degrees of freedom

Consider a Cartesian reference system centered on the point of impact and oriented so that the X-axis coincides with the normal direction and the Y-axis with the direction tangential to the impact plane, as shown in Fig. 2.8. The axes n and t coincide then with the axes X and Y, respectively.

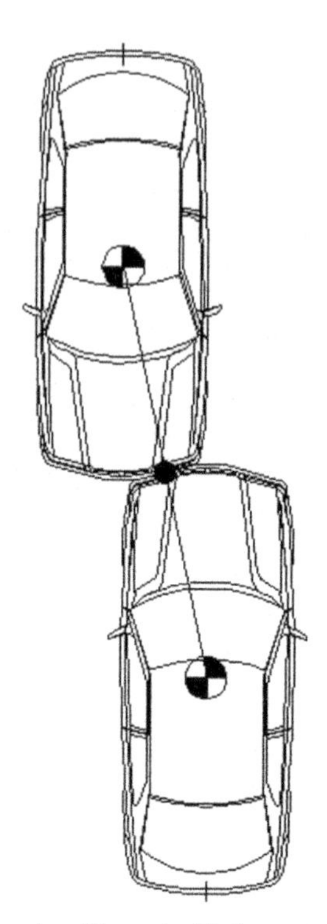

FIGURE 2.6 Alignment of the center of impact with the centers of gravity of the vehicles.

The centers of gravity of the vehicles have x and y coordinates relative to the reference system. The equations derived from the impulse theorem can be written for the two vehicles as

$$
\begin{aligned}
I_n &= m_1(\overline{V}_{1n} - V_{1n}) \\
I_t &= m_1(\overline{V}_{1t} - V_{1t}) \\
-I_n &= m_2(\overline{V}_{2n} - V_{2n}) \\
-I_t &= m_2(\overline{V}_{2t} - V_{2t})
\end{aligned}
\tag{2.20}
$$

where V_n and V_t mean components along the n and t of the centers of gravity of the vehicle velocity. The contact force, being applied in the point of impact, generates an impulsive moment that varies the moment of momentum of the two vehicles as follows:

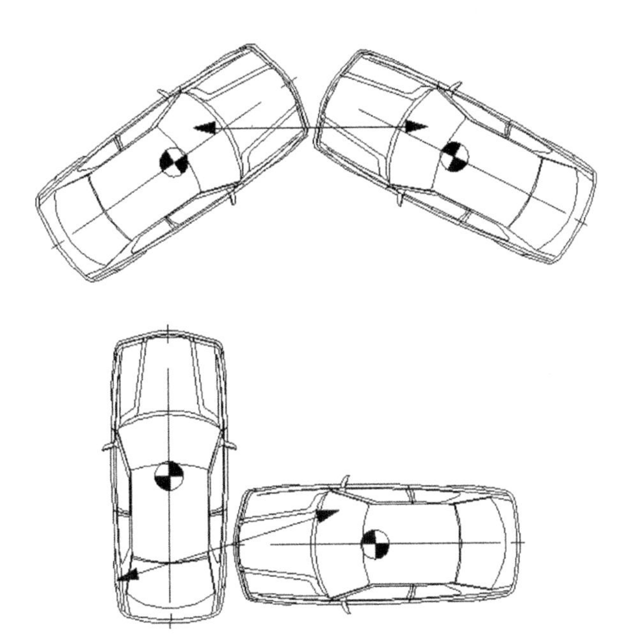

FIGURE 2.7 Oblique impacts, in which the impulse line of action does not pass through the centers of gravity of both vehicles.

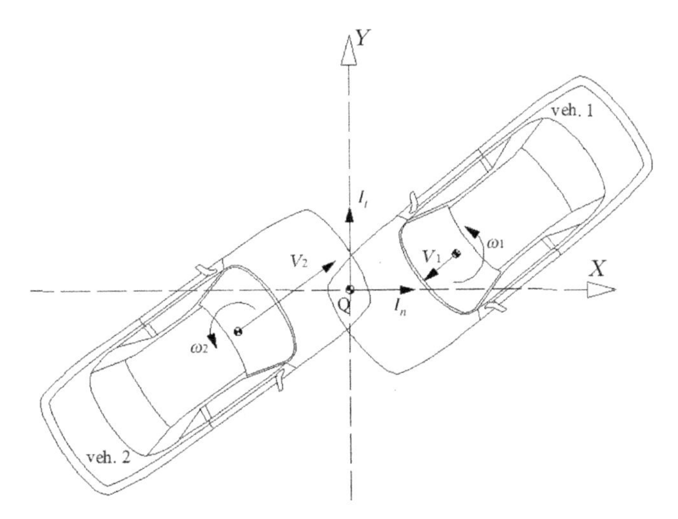

FIGURE 2.8 Diagram of the two vehicles at impact.

$$\begin{aligned} I_n y_1 - I_t x_1 &= J_1(\overline{\omega}_1 - \omega_1) \\ I_t x_2 - I_n y_2 &= J_2(\overline{\omega}_2 - \omega_2) \end{aligned} \tag{2.21}$$

where J indicates the moment of inertia of the vehicle with respect to a vertical axis perpendicular to the planes $X-Y$, and ω indicates the angular speed of the vehicles.

Resolving Eqs. (2.20) and (2.21) with respect to the final speed values one can obtain

$$\begin{aligned}
\overline{V}_{1n} &= V_{1n} + \frac{I_n}{m_1} \\
\overline{V}_{1t} &= V_{1t} + \frac{I_t}{m_1} \\
\overline{\omega}_1 &= \omega_1 + \frac{I_n}{J_1}y_1 - \frac{I_t}{J_1}x_1 \\
\overline{V}_{2n} &= V_{2n} - \frac{I_n}{m_2} \\
\overline{V}_{2t} &= V_{2t} - \frac{I_t}{m_2} \\
\overline{\omega}_2 &= \omega_2 - \frac{I_n}{J_2}y_2 + \frac{I_t}{J_2}x_2
\end{aligned} \tag{2.22}$$

from which, knowing the initial speed values, I_n and I_t, the final unknown values can be derived; and conversely, knowing the final speed values, the initial unknown values can be derived.

2.6.1 Full and sliding impact

To solve Eq. (2.22), it is necessary to determine the values of I_n and I_t through two other equations. The equations can be obtained from the closing speed definition. The normal components of the closing speed can be calculated considering the center of impact belonging alternately to the vehicle 1 and vehicle 2:

$$\begin{aligned}
\overline{V}_{Rn} &= (\overline{V}_{2n} + \overline{\omega}_2 y_2) - (\overline{V}_{1n} + \overline{\omega}_1 y_1) \\
V_{Rn} &= (V_{2n} + \omega_2 y_2) - (V_{1n} + \omega_1 y_1)
\end{aligned} \tag{2.23}$$

Analogously for the tangential component of the closing speed:

$$\begin{aligned}
\overline{V}_{Rt} &= (\overline{V}_{2t} - \overline{\omega}_2 x_2) - (\overline{V}_{1t} - \overline{\omega}_1 x_1) \\
V_{Rt} &= (V_{2t} - \omega_2 x_2) - (V_{1t} - \omega_1 x_1)
\end{aligned} \tag{2.24}$$

Substituting Eq. (2.22) into Eqs. (2.23) and (2.24), we get

$$\begin{aligned}
\overline{V}_{Rt} &= V_{Rt} + c_2 I_n - c_3 I_t \\
\overline{V}_{Rn} &= V_{Rn} + c_2 I_t - c_1 I_n
\end{aligned} \tag{2.25}$$

with

$$c_1 = \left[\frac{1}{m_1} + \frac{1}{m_2} + \frac{{y_1}^2}{J_1} + \frac{{y_2}^2}{J_2}\right]$$
$$c_2 = \left[\frac{x_1 y_1}{J_1} + \frac{x_2 y_2}{J_2}\right] \tag{2.26}$$
$$c_3 = \left[\frac{1}{m_1} + \frac{1}{m_2} + \frac{{x_1}^2}{J_1} + \frac{{x_2}^2}{J_2}\right]$$

If at the beginning of impact there is a closing speed in the tangential direction, it may occur that such speed does not vanish before the separation of the vehicles, that is, a common velocity in the center of impact is not reached, and in this case, we have a sliding impact.

In the case of full impact the closing speed in the tangential direction between vehicles, following the actions of the tangential forces, decreases until it reaches zero during the contact, which is before the impact ceases. In this case the final closing tangential speed is null.

It is possible to make a simplifying assumption, namely, to consider that at the end of the compression phase the speed of the center of impact of the two vehicles reaches a common value, for both the normal and tangential components, that is, the relative speeds became null. Such behavior may be a good approximation in most of the impacts. With this assumption, therefore, the closing speeds at the end of the compression phase in Eq. (2.25) can be put to zero and, recalling Eq. (2.17), we derive the pulse components:

$$I_n = (1 + \varepsilon_i)\frac{V_{Rn}c_3 + V_{Rt}c_2}{c_3 c_1 - {c_2}^2}$$
$$I_t = (1 + \varepsilon_i)\frac{V_{Rn}c_2 + V_{Rt}c_1}{c_3 c_1 - {c_2}^2} \tag{2.27}$$

which can be used in Eq. (2.22) to obtain the unknown values of speed.

The coefficient of restitution, in this case, expresses the ratio between the initial and final closing speeds of the center of impact in a normal direction. This coefficient is referred to as ε_i, to distinguish it from the coefficient of restitution calculated as the ratio between the relative speeds of the centers of gravity of the vehicle:

$$\varepsilon_i = -\frac{\overline{V}_{Rn}}{V_{Rn}} = \frac{(\overline{V}_{1n} - \overline{\omega}_1 x_1) - (\overline{V}_{2n} - \overline{\omega}_2 x_2)}{(V_{2n} + \omega_2 y_2) - (V_{1n} + \omega_1 y_1)} \tag{2.28}$$

It is observed that by adopting a three-degrees-of-freedom model, the coefficient ε for oblique impacts may also be negative and, in general, can be challenging to estimate a priori. Instead, the coefficient ε_i is generally between 0 and 1, except in the case of structural failure such as to divide the vehicle, and has the trend shown in Fig. 2.4.

In the case of sliding impact the tangential closing speed does not cancel during the collision, and it becomes crucial for the choice of the contact plane. The impulse direction is limited by the coefficient of friction μ between vehicles and from Eq. (2.9) one can put

$$I_t = \mu I_n \tag{2.29}$$

while for the normal impulse component, from the second of Eq. (2.25) and introducing the coefficient of restitution, it results in

$$I_n = (1 + \varepsilon_i)\frac{V_{Rn}}{c_1 - \mu c_2} \tag{2.30}$$

If from Eq. (2.27) $I_t < \mu I_n$, then there is no tangential sliding between the vehicles, the collision is full, and Eq. (2.27) can be used in Eq. (2.22) to obtain the unknown speeds. If vice versa from Eq. (2.27), we have $I_t > \mu I_n$, then we have a sliding impact, and the impulse components to be used in Eq. (2.22) are given by Eqs. (2.29) and (2.30).

The selection of the appropriate value of μ can be made by estimating a value of the PDOF, based on the deformations of the vehicles and of the impact configuration. If the values of the initial speed of the vehicles (forward calculation) are known, then by Eq. (2.8), as the final closing speed generally has a much smaller value than that of the initial one, we can identify with good approximation the direction of the impulse with the direction of initial closing speed.

Schematizing the vehicles as material points, the direction of the resultant of the force coincides with the direction of the relative speed. In fact, $c_2 = 0$, $c_1 = c_3 = (1/m_1 + 1/m_2)$ and then $I_t/I_n = V_{Rt}/V_{Rn}$.

In the sliding impacts the value of μ has, however, a maximum value, beyond which the relative sliding in the tangential direction ceases before the separation of the vehicles, and the impact becomes full. The maximum value of the parameter μ is determined in Section 2.8.2.

Eqs. (2.27) and (2.30) can be derived using only terms relating to final conditions, to be used in a backward analysis:

$$\begin{aligned} I_n &= -\frac{(1 + \varepsilon_i)}{\varepsilon_i}\frac{\overline{V}_{Rn}c_3 + \overline{V}_{Rt}c_2}{c_3c_1 - c_2{}^2} \\ I_t &= -\frac{(1 + \varepsilon_i)}{\varepsilon_i}\frac{\overline{V}_{Rn}c_2 + \overline{V}_{Rt}c_1}{c_3c_1 - c_2{}^2} \end{aligned} \tag{2.31}$$

$$I_n = -\frac{(1 + \varepsilon_i)}{\varepsilon_i}\frac{\overline{V}_{Rn}}{c_1 - \mu c_2} \tag{2.32}$$

It is to be highlighted that Eqs. (2.31) and (2.32) are hill conditioned, as the restitution coefficient is on the denominator, and little errors in their value produce a large variation on the impulses.

To derive the unknown speeds the eight equations in Eqs. (2.20), (2.21), and (2.25) have been solved. It is convenient to write the resolutive equations in matrix form for compactness and practicality of implementation of the expressions in a calculation software. It is also convenient to eliminate from the system the two unknowns I_n and I_t, to get a system in which only the six unknown speeds are displayed.

Eliminating the terms I_n and I_t from Eq. (2.20), we obtain the general equations expressing the conservation of momentum in both directions n and t:

$$\begin{aligned} m_1\overline{V}_{1n} + m_2\overline{V}_{2n} &= m_1 V_{1n} + m_2 V_{2n} \\ m_1\overline{V}_{1t} + m_2\overline{V}_{2t} &= m_1 V_{1t} + m_2 V_{2t} \end{aligned} \tag{2.33}$$

Similarly, substituting the expressions of I_n and I_t, obtainable by Eq. (2.20) in Eq. (2.21), we get

$$\begin{aligned} m_1y_1\overline{V}_{1n} - m_1y_1V_{1n} - m_1x_1\overline{V}_{1t} + m_1x_1V_{1t} &= J_1(\overline{\omega}_1 - \omega_1) \\ m_2y_2\overline{V}_{2n} - m_2y_2V_{2n} - m_2x_2\overline{V}_{2t} + m_2x_2V_{2t} &= J_2(\overline{\omega}_2 - \omega_2) \end{aligned} \tag{2.34}$$

For full impacts the condition that the final tangential closing speed is zero can be assumed, namely,

$$\overline{V}_{Rt} = (\overline{V}_{2t} - \overline{\omega}_2x_2) - (\overline{V}_{1t} - \overline{\omega}_1x_1) = 0 \tag{2.35}$$

whereas for the sliding impacts Eq. (2.29) is

$$m_1\overline{V}_{1t} - m_1V_{1t} = \mu m_1\overline{V}_{1n} - \mu m_1V_{1n} \tag{2.36}$$

Eqs. (2.33), (2.34), (2.28), and (2.35) or (2.36) constitute the searching equations in six unknown speeds, which in matrix form can be written, in that case Eq. (2.35) is used for full impacts:

$$\begin{bmatrix} m_1 & 0 & 0 & m_2 & 0 & 0 \\ 0 & m_1 & 0 & 0 & m_2 & 0 \\ y_1m_1 & -x_1m_1 & -J_1 & 0 & 0 & 0 \\ 0 & 0 & 0 & y_2m_2 & -x_2m_2 & -J_2 \\ -1 & 0 & -y_1 & 1 & 0 & y_2 \\ 0 & -1 & -x_1 & 0 & 1 & -x_2 \end{bmatrix} \begin{Bmatrix} \overline{V}_{1n} \\ \overline{V}_{1t} \\ \overline{\omega}_1 \\ \overline{V}_{2n} \\ \overline{V}_{2t} \\ \overline{\omega}_2 \end{Bmatrix} = \begin{bmatrix} m_1 & 0 & 0 & m_2 & 0 & 0 \\ 0 & m_1 & 0 & 0 & m_2 & 0 \\ y_1m_1 & -x_1m_1 & -J_1 & 0 & 0 & 0 \\ 0 & 0 & 0 & y_2m_2 & -x_2m_2 & -J_2 \\ \varepsilon_i & 0 & \varepsilon_iy_1 & -\varepsilon_i & 0 & -\varepsilon_iy_2 \\ 0 & 0 & 0 & 0 & 0 & 0 \end{bmatrix} \begin{Bmatrix} V_{1n} \\ V_{1t} \\ \omega_1 \\ V_{2n} \\ V_{2t} \\ \omega_2 \end{Bmatrix} \tag{2.37}$$

alternatively, in the case of using Eq. (2.36) for sliding impacts:

$$\begin{bmatrix} m_1 & 0 & 0 & m_2 & 0 & 0 \\ 0 & m_1 & 0 & 0 & m_2 & 0 \\ y_1 m_1 & -x_1 m_1 & -J_1 & 0 & 0 & 0 \\ 0 & 0 & 0 & y_2 m_2 & -x_2 m_2 & -J_2 \\ -1 & 0 & -y_1 & 1 & 0 & y_2 \\ -\mu m_1 & m_1 & 0 & 0 & 0 & 0 \end{bmatrix} \begin{Bmatrix} \overline{V}_{1n} \\ \overline{V}_{1t} \\ \overline{\omega}_1 \\ \overline{V}_{2n} \\ \overline{V}_{2t} \\ \overline{\omega}_2 \end{Bmatrix} = \begin{bmatrix} m_1 & 0 & 0 & m_2 & 0 & 0 \\ 0 & m_1 & 0 & 0 & m_2 & 0 \\ y_1 m_1 & -x_1 m_1 & -J_1 & 0 & 0 & 0 \\ 0 & 0 & 0 & y_2 m_2 & -x_2 m_2 & -J_2 \\ \varepsilon_i & 0 & \varepsilon_i y_1 & -\varepsilon_i & 0 & -\varepsilon_i y_2 \\ -\mu m_1 & m_1 & 0 & 0 & 0 & 0 \end{bmatrix} \begin{Bmatrix} V_{1n} \\ V_{1t} \\ \omega_1 \\ V_{2n} \\ V_{2t} \\ \omega_2 \end{Bmatrix} \quad (2.38)$$

In symbolic notation, we can write

$$[\overline{\mathbf{M}}]\, \overline{\mathbf{V}} = [\mathbf{M}]\, \mathbf{V} \quad (2.39)$$

from which the direct solution (forward calculation) is obtained, which provides the final speed, knowing the initial ones:

$$\overline{\mathbf{V}} = [\overline{\mathbf{M}}]^{-1} [\mathbf{M}]\, \mathbf{V} = [\mathbf{F}]\, \mathbf{V} \quad (2.40)$$

or the inverse solution (backward calculation), which provides the initial speeds knowing the final ones:

$$\mathbf{V} = [\mathbf{M}]^{-1} [\overline{\mathbf{M}}]\, \mathbf{V} = [\mathbf{B}]\, \overline{\mathbf{V}} \quad (2.41)$$

The matrix [**M**], for the restitution coefficient values ε_ι equal to zero, is singular, that is not invertible, and then the solution cannot be found. This is detectable also from Eqs. (2.31) and (2.32), in which the ε_ι coefficient appears in the denominator. In these cases a minimal value can be assigned to ε_ι, and the forward solution can be used by assigning attempt values to the initial speeds, gradually adjusting the values until the final data coincide with those known, by an accepted error.

2.6.2 Coefficient of restitution at the center of impact

The relationship between the coefficient of restitution referred to the center of mass of vehicles, and that referred to the impact center is here obtained. For simplicity, but without loss of generality, reference is made to a reference system aligned with the PDOFs, as shown in Fig. 2.9.

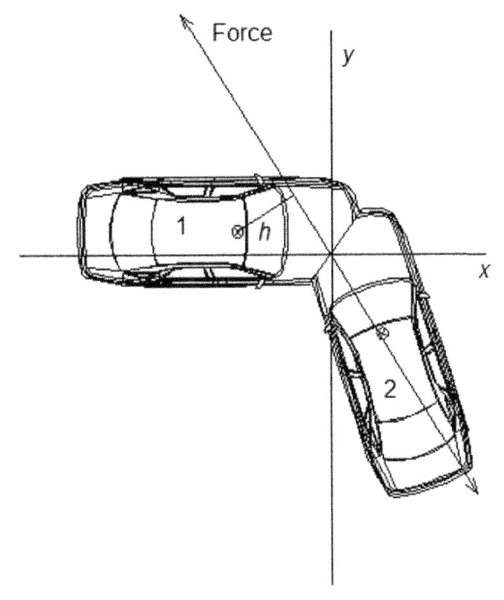

FIGURE 2.9 Diagram of the vehicles at the time of the maximum exchange of forces. Both the direction of the resultant of the contact force and its arm h with respect to the center of gravity of the vehicle 1 is highlighted.

The coefficient of restitution referred to the impact center ε_i is defined as

$$\varepsilon_i = \frac{\overline{u}_1 - \overline{u}_2}{u_2 - u_2} = \frac{\overline{V}_{1\mathrm{PDOF}} + h_1\overline{\omega}_1 - \overline{V}_{2\mathrm{PDOF}} - h_2\overline{\omega}_2}{V_{2\mathrm{PDOF}} + h_2\omega_2 - V_{1\mathrm{PDOF}} - h_1\omega_1} \tag{2.42}$$

where u and V_{PDOF} are the components along the direction of PDOF of the velocity of the center of impact and center of gravity, respectively; ω are the angular velocities of vehicles; and h the arms of the force resultant referred to the center of gravity.

From Eq. (2.42), collecting the terms and recalling that $\overline{V}_{\mathrm{PDOF}} - V_{\mathrm{PDOF}} = \Delta V$ and $\overline{\omega} - \omega = \Delta\omega$, with a few steps it is obtained that

$$(1 + \varepsilon_i) = \frac{\Delta V_1 - \Delta V_2 + h_1\Delta\omega_1 - h_2\Delta\omega_2}{V_{2\mathrm{PDOF}} + h_2\omega_2 - V_{1\mathrm{PDOF}} - h_1\omega_1} \tag{2.43}$$

As $Ih = m\Delta Vh = J\Delta\omega$, obtaining $\Delta\omega$ and replacing in Eq. (2.43), with some calculations, we obtain the following equation i:

$$(1 + \varepsilon_i) = \frac{\Delta V_1/\gamma_1 - \Delta V_2/\gamma_2}{V_{2\mathrm{PDOF}} + h_2\omega_2 - V_{1\mathrm{PDOF}} - h_1\omega_1} \tag{2.44}$$

where γ is the factor of mass reduction that is correlated to the impact eccentricity for the given vehicle, expressed by

$$\gamma = \frac{k^2}{k^2 + h^2} \tag{2.45}$$

with k radius of gyration of the vehicle: $k = J/m$ (J indicates the moment of inertia of vehicle around the vertical axis) and h is arm of the resultant of forces (see Fig. 2.9).

If we assume that the rotational velocity of both vehicles before the impact is negligible, Eq. (2.44) becomes

$$(1+\varepsilon_i) = \frac{\Delta V_1/\gamma_1 - \Delta V_2/\gamma_2}{V_{2\mathrm{PDOF}} - V_{1\mathrm{PDOF}}} = \frac{\Delta V_1/\gamma_1 - \Delta V_2/\gamma_2}{V_{c\mathrm{PDOF}}} \tag{2.46}$$

For the coefficient of restitution referred to the center of mass of vehicles, it is possible to write

$$(1+\varepsilon) = \frac{V_{2\mathrm{PDOF}} - V_{1\mathrm{PDOF}} + \overline{V}_{1\mathrm{PDOF}} - \overline{V}_{21\mathrm{PDOF}}}{V_{2\mathrm{PDOF}} - V_{1\mathrm{PDOF}}} = \frac{\Delta V_1 - \Delta V_2}{V_{c\mathrm{PDOF}}} \tag{2.47}$$

By dividing Eq. (2.47) by Eq. (2.46)

$$\frac{(1+\varepsilon)}{(1+\varepsilon_i)} = \frac{\gamma_1\gamma_2(\Delta V_1 - \Delta V_2)}{\Delta V_1\gamma_2 - \Delta V_2\gamma_1} \tag{2.48}$$

As for any impact $\Delta V_2 = -\Delta V_1 m_1/m_2$, by replacing it in the previous equation and simplifying

$$\frac{(1+\varepsilon)}{(1+\varepsilon_i)} = \frac{\gamma_2\gamma_1(m_1+m_2)}{m_2\gamma_2 + m_1\gamma_1} \tag{2.49}$$

or

$$\frac{(1+\varepsilon)}{(1+\varepsilon_i)} = \frac{m_c^*}{m_c} \tag{2.50}$$

where m_c^* is the common mass calculated with reduced masses $m\gamma$:

$$m_c^* = \frac{m_1\gamma_1 m_2\gamma_2}{m_1\gamma_1 + m_2\gamma_2} \tag{2.51}$$

This relationship, obtained for a reference system aligned with the PDOF, is valid for any reference system.

2.6.3 Directions of pre- and postimpact speeds

By imposing the conservation of momentum, four equations were written, which link the initial speeds of the vehicles to the final ones. To calculate the speeds, other two equations have been added, imposing a proportionality relationship between initial and final closing speeds. Such proportionality has been expressed through the coefficients of restitution ε_i and friction μ. The use of these coefficients constitutes an alternative way to the use of energy balance, taking account of the energy dissipated in the plastic

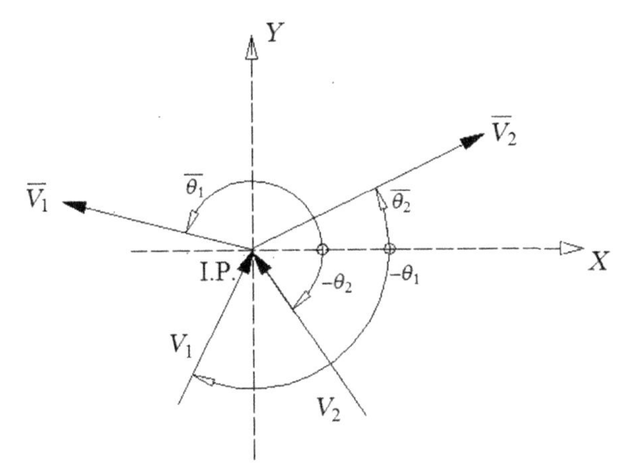

FIGURE 2.10 Speed of the vehicles before and after the collision.

deformations of the vehicles, with the advantage of obtaining a system of linear equations.

In many practical cases, by a priori information or estimates based on reliable information, directions of travel taken by the vehicles before and after the collision can be known. In this case, we have two additional information, which can be used together with equations expressing the conservation of momentum, without having to use the coefficients mentioned previously. Taking an arbitrary reference system $X-Y$ centered on the point of impact, directions of the vehicle speeds can be considered schematically shown in Fig. 2.10, and the initial speeds can be written as

$$\begin{aligned} V_{1x} &= V_1\cos(\theta_1) \\ V_{1y} &= V_1\sin(\theta_1) \\ V_{2x} &= V_2\cos(\theta_2) \\ V_{2y} &= V_2\sin(\theta_2) \end{aligned} \tag{2.52}$$

Similar relationships can be written for the final speeds of the vehicles. Substituting these relations in Eq. (2.33), written for the reference $X-Y$, the following two equations can be obtained:

$$\begin{aligned} m_1V_1\cos(\theta_1) + m_2V_2\cos(\theta_2) &= m_1\overline{V}_1\cos(\overline{\theta}_1) + m_2\overline{V}_2\cos(\overline{\theta}_2) \\ m_1V_1\sin(\theta_1) + m_2V_2\sin(\theta_2) &= m_1\overline{V}_1\sin(\overline{\theta}_1) + m_2\overline{V}_2\sin(\overline{\theta}_2) \end{aligned} \tag{2.53}$$

which, knowing the incoming impact speeds of the vehicles, θ_1 and θ_2, and the final velocities in magnitude and direction, the unknown initial speeds V_1 and V_2 can be solved:

$$V_2 = \frac{1}{(f - eb/a)}\left[\overline{V}_2(h - ed/a) + \overline{V}_1(g - ec/a)\right]$$
$$V_1 = \frac{1}{a}(d\overline{V}_2 + c\overline{V}_1 - bV_2) \tag{2.54}$$

with

$$\begin{aligned} a &= m_1\cos(\theta_1) \\ b &= m_2\cos(\theta_2) \\ c &= m_1\cos(\overline{\theta}_1) \\ d &= m_2\cos(\overline{\theta}_2) \\ e &= m_1\sin(\theta_1) \\ f &= m_2\sin(\theta_2) \\ g &= m_1\sin(\overline{\theta}_1) \\ h &= m_2\sin(\overline{\theta}_2) \end{aligned} \tag{2.55}$$

obtaining the travel speed of the vehicle, Eq. (2.34) provides the unknown angular speeds.

When handling Eq. (2.53), one must be careful to consider the momentum with their sign. The use of a coordinate system expressed with angles from 0 to 360 degrees, positive counterclockwise, as shown in Fig. 2.10 allows to consider the proper sign associated to trigonometric functions automatically.

2.6.4 Impact against a rigid barrier

One considers the diagram in Fig. 2.11, where a reference system with the Y-axis parallel to the barrier and coinciding with the t-axis has been chosen. In the illustrated configuration the components of the speed of the vehicle are negative.

By indicating with the subscript 1 the vehicle and with the subscript 2 the barrier, we have $1/m_2 = 1/J_2 = 0$, $V_{2x} = V_{2y} = \omega_2 = 0$ and, in the case of initial rotation of the vehicle null, even $\omega_1 = 0$, then

$$c_1 = \left[\frac{1}{m_1} + \frac{{y_1}^2}{J_1}\right]$$
$$c_2 = \left[\frac{x_1 y_1}{J_1}\right] \tag{2.56}$$
$$c_3 = \left[\frac{1}{m_1} + \frac{{x_1}^2}{J_1}\right]$$

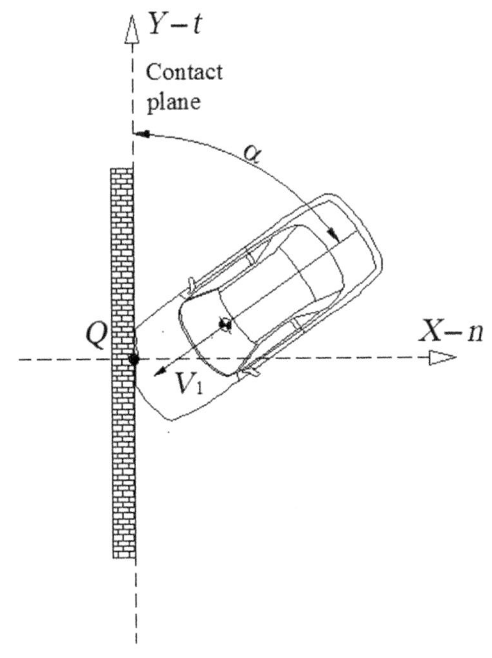

FIGURE 2.11 Angled impact against a barrier.

The initial closing speed, in the direction perpendicular to the barrier, is $V_{Rx} = -V_{1x}$, while in the tangential direction is $V_{Ry} = V_{1y}$.

If from Eq. (2.27), it is $I_t < \mu I_n$ then there is no tangential sliding, that is, the vehicle during the collision ceases the tangential sliding along the rigid barrier, Eq. (2.27) can be used in Eq. (2.22) to obtain the unknown speed values. The angular velocity is negative, that is, the rotation of the vehicle is clockwise for the configuration shown in Fig. 2.11.

If vice versa, from Eq. (2.27) results in $I_t > \mu I_n$ then we have a sliding collision, that is, the vehicle slides along the barrier and the pulse components to be used in Eq. (2.22) are given by Eqs. (2.29) and (2.30).

In this last case, you get the final speed values:

$$
\begin{aligned}
\overline{V}_{1x} &= \frac{m_1 {y_1}^2 - \mu^2 m_1 x_1 y_1 - \varepsilon_i J_1}{m_1 {y_1}^2 - \mu^2 m_1 x_1 y_1 + J_1} V_{1x} \\
\overline{V}_{1y} &= V_{1y} - \frac{J_1 \mu (1 + \varepsilon_i)}{m_1 {y_1}^2 - \mu^2 m_1 x_1 y_1 + J_1} V_{1x} \\
\overline{\omega}_1 &= \frac{m_1 (1 + \varepsilon_i)(\mu x_1 - y_1)}{m_1 {y_1}^2 - \mu^2 m_1 x_1 y_1 + J_1} V_{1x}
\end{aligned}
\tag{2.57}
$$

The sign of the final angular speed depends on the sign of $(\mu x_1 - y_1)$, that is, from the position of the point of impact relative to the vehicle center of gravity. For large values of the angle α of incidence of the vehicle, recalling that in this case V_{1x} is negative, a negative angular velocity is obtained, or a clockwise rotation in the configuration shown in Fig. 2.11. Conversely, for small angles of incidence α the vehicle swipes along the rigid barrier and rotates counterclockwise, that is, it is redirected toward the roadway.

2.7 Sensitivity analysis

The system of equations written in the form Eq. (2.41) highlights how the matrix [**B**] may be considered a kind of transfer function of the system constituted by the two vehicles at the moment of impact. The matrix, that is, determines the terms and the amount by which the unknown values are obtained from the known values of speed.

In the matrix [**B**] does not appear the speed values, which are grouped in the input and output vectors. The matrix contains the geometric elements (positions of the center of the impulse), the characteristics of the vehicles (the masses and moments of inertia), and the characteristics of impact (coefficients of restitution and μ) instead.

The matrix [**B**], therefore, defines the "type" of a collision and its configuration and characterizes the system's response, regardless of the vehicle speed values.

A useful parameter for estimating the effect of amplification of uncertainties in the initial data on the final data is the conditioning number $\chi(B)$ of the [**B**] matrix, which furnishes an index of global sensitivity to errors in calculating the unknown velocity values. As a result,

$$\frac{\|\Delta\mathbf{V}\|}{\|\mathbf{V}\|} \leq \chi(B)\frac{\|\Delta\overline{\mathbf{V}}\|}{\|\overline{\mathbf{V}}\|} \tag{2.58}$$

where $\|\mathbf{V}\|$ indicates the norm of vector $\mathbf{V}$, $\Delta\mathbf{V}$ is the speed errors. It results in $\chi(B) \geq 1$. This is an a priori indication that the relative error in $\mathbf{V}$ is at most equivalent to $\chi(B)$ times the relative error in $\overline{\mathbf{V}}$. High sensitivity to errors, typical of an ill-conditioned matrix, furnishes a high value of the conditioning number, which translates into pronounced instability in the system; a minor variation in the input data results in a significant variation in the output data. In the presence of a type of collision characterized by an ill-conditioned [**B**] matrix, a range, even limited, of uncertainty in the post-collision velocity values may provide several precollision velocity scenarios differing widely from one another, all congruent with the uncertainty in the initial data. A useful parameter in this context is the norm of matrix [**B**]. In fact,

$$\|\Delta\mathbf{V}\| \leq \|\mathbf{B}\|\,\|\Delta\overline{\mathbf{V}}\| \tag{2.59}$$

The norm is thus linked directly to the absolute errors in velocity.

If the backward calculation is unstable, with a high value of the norm, by inverting the calculation procedure, utilizing Eq. (2.40) instead of Eq. (2.41) and then hypothesizing that the precollision velocity values are known, it is possible to obtain a more stable system, with a low value of the [**F**] matrix norm. In this case, by varying the input data over a still broader uncertainty range, limited uncertainty ranges in the postcollision velocity values may be obtained.

However, this situation, which of knowing the precollision velocities, is a case that rarely occurs in reconstructing traffic accidents. In general, the unknown velocities are the precollision ones. In using forward calculation the procedure is that of assigning tentative precollision velocities until finding, by successive iterations, postcollision values that coincide with the known postcollision ones. It follows that the sensitivity to errors in the input data is the same, even though the calculation procedure is inverted. The significant parameter for evaluating sensitivity is thus usually the [**B**] matrix norm.

To obtain more detailed information on the system's sensitivity to error and to determine which parameters are the most critical, the elements of the [**B**] matrix may be analyzed, grouped by rows, columns, or submatrices. The individual elements in the [**B**] matrix constitute the coefficients of influence of the system. For example, the element in the [**B**] $(n \times m)$ matrix at row k and column i, B_{kl}, expresses the influence, or the amplification, of the term $\overline{V}_i$ on the calculated value of V_k, which is

$$V_k = B_{k1}\overline{V}_l + B_{k2}\overline{V}_2 + \cdots + B_{kn}\overline{V}_n \tag{2.60}$$

By analyzing the values of the elements in the [**B**] matrix, it is thus possible to evaluate the weight of the individual known velocity values in calculating the unknown individual values. Given the uncertainty ranges $\mathbf{\Delta\overline{V}}$ on postcollision velocities, on V_k there is an uncertainty range of

$$\Delta V_k = \sum_{l=1}^{n} B_{kl}\Delta\overline{V}_l \tag{2.61}$$

We assume that the width of the $\mathbf{\Delta\overline{V}}$ uncertainty is equal for all the components of velocity. In this case the 1-norm N_k of row k, defined as

$$N_k = \sum_{l=1}^{n} \left|\underline{B}_{kl}\right| \tag{2.62}$$

furnishes a higher value of uncertainty amplification for all of the input data employed in calculating Vl_k, regardless of the value of the uncertainties. It thus furnishes an a priori estimation of sensitivity to error. This sensitivity to errors is typical of the "type" of accident considered, regardless of the velocity values.

If the width of the $\Delta\overline{\mathbf{V}}$ uncertainty is not equal for all the speed components, it is possible to consider a weighted sum of the terms of each row, with the weights given by the uncertainty intervals themselves.

$$N_k = \sum_{l=1}^{n} \left|B_{kl}\Delta\overline{V}_l\right| \tag{2.63}$$

Analysis by rows makes it possible to determine which unknown value is most sensitive to errors in the input data and allows the analyst to attribute a just weight to the calculated values, taking account of their reliability.

Conversely, the norm of the columns in the [**B**] matrix can be analyzed:

$$N_l = \sum_{k=1}^{m} \left|\underline{B}_{kl}\right| \tag{2.64}$$

thus obtaining an indication of how much an error in an individual postcollision velocity value influences the calculation of the unknown values, that is, allows to assess the criticality of an input value. This indication is highly useful to the analyst since, in the case of high criticality of an input data (high value of the norm for the corresponding column), there is a suggestion that the datum in question should be more thoroughly evaluated, for instance through a more accurate analysis of the postcollision stage, in order to minimize its uncertainty.

Example

The case of one vehicle running into another may be considered, analyzed with a model having one degree of freedom and assuming the following data, determined through previous analyses:

$$\begin{aligned} m_a &= 2000\ \text{kg} \\ m_b &= 800\ \text{kg} \\ \varepsilon &= 0.3 \\ \overline{V}_a &= 8\ \text{km}/h \\ \overline{V}_b &= 12\ \text{km}/h \end{aligned}$$

The conservation of momentum, together with a definition of the coefficient of restitution, written in matrix form is

$$\begin{bmatrix} m_a & m_b \\ -\varepsilon & \varepsilon \end{bmatrix} \begin{Bmatrix} V_a \\ V_b \end{Bmatrix} = \begin{bmatrix} m_a & m_b \\ 1 & -1 \end{bmatrix} \begin{Bmatrix} \overline{V}_a \\ \overline{V}_b \end{Bmatrix}$$

by Eq. (2.41), we obtain a matrix [**B**]:

$$[\mathbf{B}] = \begin{bmatrix} -0.2 & 1.2 \\ 3.1 & -2.1 \end{bmatrix}$$

with the following characteristics:

matrix norm = 3.7
row 1 norm 1 = 1.5
row 2 norm = 5.2
column 1 norm 1 = 3.3
column 2 norm = 3.3

The norm value indicates an overall sensitivity of 3.7 compared to the error on the initial data. So whether on the latter there is an uncertainty of ± 3 km/h, on the final results can be expected uncertainty of about $\pm 3 \times 3.7 = 11$ km/h.

This uncertainty value is mediated between the two unknown speeds, V_a and V_b, but this does not mean that for both unknown speeds, the same sensitivity to error exists. More detailed information is obtained by analyzing the norm for rows and columns.

The value of the norm of row 1 indicates that the sensitivity of the calculation for the speed V_a, (1.5) is much lower than that for the V_b speed (52). The calculation of the initial speed of the vehicle b is, therefore, more delicate, showing an uncertainty of $\pm 3 \times 5.2 = 15.6$ km/h against to uncertainty of $\pm 3 \times 1.5 = 4.5$ km/h for the vehicle a.

It can be seen, however, that the input data of the two vehicles similarly affect the results, the norm of their respective columns being equal for both.

This analysis can be repeated for different values of ε. Fig. 2.12 shows the trend of the value of the norm of the matrix [**B**] as a function of the value of ε. Note how the decrease of ε increases the sensitivity of the calculation error.

To entirely plastic impact the matrix inversion shown in Eq. (2.41) is not possible, and the solution is indeterminate.

In the case of conservation of momentum models with three degrees of freedom, the f [B] matrix is (6×6). In this case an useful analysis is that of

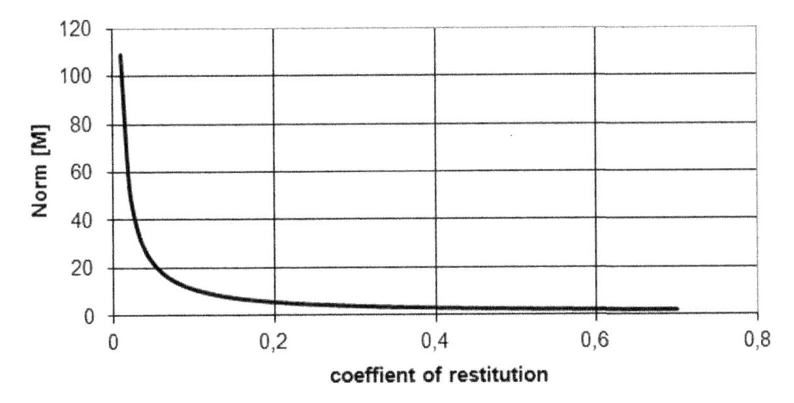

FIGURE 2.12 Graph of [M] norm as a function of the coefficient of restitution ε.

determining the norm of the submatrices (3 × 3), **X**, **Y**, **H**, and **K**, obtained as

$$\begin{Bmatrix} V_A \\ V_B \end{Bmatrix} = \begin{bmatrix} \mathbf{X} & \mathbf{H} \\ \mathbf{K} & \mathbf{Y} \end{bmatrix} \begin{Bmatrix} \overline{V}_A \\ \overline{V}_B \end{Bmatrix} \tag{2.65}$$

where $\overline{V}_a = \{\overline{u}_a, \overline{v}_a, \overline{\omega}_a\}$ $\overline{V}_b = \{\overline{u}_b, \overline{v}_b, \overline{\omega}_b\}$ and are the vectors of the postcollision velocities relevant to vehicles a and b respectively, and, by analogy, the vectors V_a and V_b refer to the precollision velocities.

The submatrices [**X**] and [**H**] determine the influence of the postcollision velocities of a given vehicle on the precollision velocities of the same vehicle. The submatrices [**B**] and [**C**] instead determine the influence of the postcollision velocities of a given vehicle on the precollision velocities of the other vehicle.

The norms of the submatrices thus furnish useful information on how much the uncertainties on the data obtained, for instance, by postcollision analysis for a given vehicle are correlated to the unknown velocities of the two vehicles. In some cases the uncertainty values on the postcollision velocities of one vehicle may be more significant than on those of the other. By analyzing the norms of the submatrices, it is possible to show which vehicle is most severely affected by the uncertainty or, vice versa, which calculated data are the most reliable.

Example

Consider, for example, a similar impact to that of example preceding, but the two vehicles do not collide in a perfectly centered way, that is, the lines of action of the impulses do not pass precisely to the centers of gravity of the vehicles. Consider an offset of 0.8 m, as shown in Fig. 2.13. We use an impact model in three degrees of freedom.

The following [**B**] matrix is obtained:

$$[\mathbf{B}] = \begin{bmatrix} -0.2 & 0.4 & 1.2 & 1.2 & -0.4 & 1.2 \\ -0.2 & 1.9 & 1.6 & 0.2 & -0.9 & 1.6 \\ 0.1 & 0.9 & 2.7 & -0.1 & -0.9 & 1.7 \\ 3.1 & -0.9 & -2.9 & -2.1 & 0.9 & -2.9 \\ 0.5 & -2.2 & -4.1 & -0.5 & 3.2 & -4.1 \\ 0.3 & 2.4 & 4.1 & -0.3 & -2.4 & 5.1 \end{bmatrix}$$

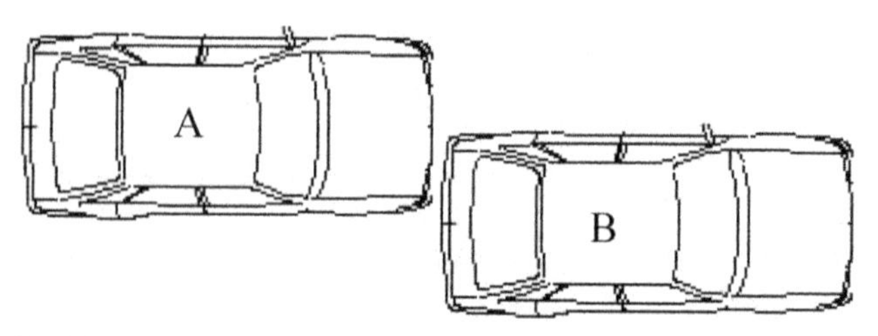

FIGURE 2.13 Diagram showing the location of the vehicles at impact.

having the following characteristics:

Rows norm 1 = 4	Columns norm 1 = 4	Submatrix norm $A = 3.8$
Rows norm 2 = 6	Columns norm 2 = 9	Submatrix norm $B = 3.0$
Rows norm 3 = 6	Columns norm 3 = 17	Submatrix norm $C = 7.5$
Rows norm 4 = 13	Columns norm 4 = 4	Submatrix norm $D = 8.2$
Rows norm 5 = 14	Columns norm 5 = 9	
Rows norm 6 = 14	Columns norm 6 = 17	Matrix norm $M = 12$

Analyzing the values of the coefficients of influence, that is, the terms of the matrix [**M**], one sees how the terms (1.1), (1.4), (4.1), and (4.4) relative to the values of the speeds along the longitudinal directions of the vehicles were unchanged compared to the previous example. This indicates an equal sensitivity to errors while appearing all other terms of the matrix, which reveal information about the sensitivity to errors due to other components of velocity, the transverse and rotation eventually held by the vehicles.

From the analysis of the norms, it is evident that also in this case the calculation of the speed of the vehicle B is more sensitive to initial uncertainties, and in particular to the uncertainties on the final values of the rotation speed of both vehicles and lateral translation. The presence and extent of such velocity components at the end of the impact must then be carefully estimated. In fact, an uncertainty of ± 1 rad/s on a car's rotation speed may lead to an error ± 2.9 km/h on the longitudinal speed of the vehicle B and an error ± 4.1 rad/s on the rotation speed of the vehicle B before impact and ± 2.7 rad/s of the vehicle A.

This uncertainty can nullify a subsequent analysis, for example, to establish an evasive steering maneuver carried out by one of the two vehicles.

Fig. 2.14 shows the modules of the traveling speed obtained by simulation with Monte Carlo, taking as initial speeds uncertainty the interval

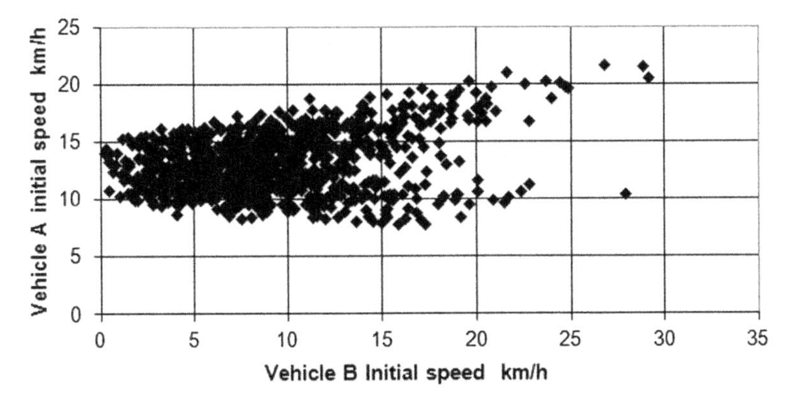

FIGURE 2.14 Variation range of vehicle speeds before impact, evaluated with Monte Carlo.

± 3 km/h for translational components and ± 1 rad/s for rotational components, all with a uniform probability distribution.

It can be seen how the range of variation found substantially coincides with the maximum expected error from the analysis of the norms, considering the weighted sums of the rows and calculating the module of the traveling speed by the longitudinal and transverse components. As shown in the earlier example, the analyses of the structure of the matrix [**M**] and the Monte Carlo simulation provide similar information on the spread or uncertainty of results.

However it should be noted that this result is obtained with Monte Carlo once set the postimpact velocity values, while the analysis of the norms determines the sensitivity to error regardless of the input values, depending only on the "type" of impact under consideration, that is, from the geometry and characteristics of the vehicles and impact.

Also with Monte Carlo, there is no indication about the criticality of the input parameters, while the analysis of the norms and coefficients of influence, providing just these indications, helps improve the calculation result, acquiring, for example, details relating to the postimpact phase to reduce the uncertainty of the most critical input data.

2.8 Energy

The kinetic energy of the vehicles, schematized as rigid bodies moving in the plane, is due to the travel speed of the centers of gravity and their angular velocities along the vertical axis. At the initial time can be written

$$E = \frac{1}{2}m_1\left(V_{1x}^2 + V_{1y}^2\right) + \frac{1}{2}m_2\left(V_{2x}^2 + V_{2y}^2\right) + \frac{1}{2}J_1\omega_1^2 + \frac{1}{2}J_1\omega_1^2 \tag{2.66}$$

A similar equation for the kinetic energy at the end of impact can be written, replacing the initial quantities with those finals. In each collision a portion of the kinetic energy initially possessed by the vehicle is converted into other forms of energy, in part is transferred from one vehicle to another and in part remains as the residual kinetic energy of each vehicle. One can write the energy dissipated as

$$E_d = E - \overline{E} \tag{2.67}$$

In the calculation of the energies involved, neglecting the rotational kinetic energy of the vehicles leads to a mistake in estimating the energy dissipated. In most practical cases the initial rotational kinetic energy of the vehicles is negligible. In these cases, by performing a calculation of the speeds with a model with two degrees of freedom and thus neglecting also the final rotational energy, an overestimation of the dissipated energy is always obtained.

2.8.1 Energy and Kelvin's theorem

Consider two vehicles schematized as point masses, with two degrees of freedom. The kinetic energy of the system can be expressed at any time as a sum of kinetic energy possessed by the center of gravity plus the kinetic energy due to the closing speed between the two vehicles:

$$E = \frac{1}{2}(m_1 + m_2)\mathbf{V}_{\mathbf{G}}^{2} + \frac{1}{2}m_c\mathbf{V}_{\mathbf{R}}^{2} \tag{2.68}$$

with

$$m_c = \frac{m_1 m_2}{m_1 + m_2} \tag{2.69}$$

Since the center of gravity speed is constant at every instant, the change in kinetic energy of the system is due solely to the variations of the closing speed. The latter, if energy from outside is not supplied as in practice takes place, is maximum at the beginning of impact, is canceled at the time of the end of the compression phase and regains, at the end of the collision, a value, with the sign reversed. The magnitude of the speed regained depends on the extent of the elastic restitution, always resulting in less or equal to the initial one.

It is useful, for an energy balance and to analyze the mode of energy transfer from one vehicle to another, splitting the collision into two steps: the step of compression and that of restitution.

Compression phase

At this stage the vehicles are in contact with each other with an initial closing speed V_R; during the contact, forces have generated that change the momentum of each vehicle, while remaining constant, as seen, its total value. In this contact the centers of gravity of the vehicles approach one another, due to the deformation of the structures, up to reaching a position of minimum distance, which corresponds to a zero closing speed. The common speed reached by the vehicles then coincides with the speed $\mathbf{V_G}$ of the center of gravity of the system, and the kinetic energy of the system reaches its minimum value.

At this stage a part of the initial kinetic energy possessed by the vehicles is absorbed in the deformation of the vehicles. This absorbed energy E_a has two components: elastic and inelastic. The elastic component E_r is retrieved from the system in the next phase of restitution, while the inelastic one is associated mainly to permanent deformation of the structures. Therefore in the following, we use indifferently the term plastic or inelastic and also to the viscous behavior of the material and the conversion to the other forms of energy such as acoustic energy, vibration, and heat. The inelastic component, for the energy balance, it is considered dissipated energy E_d, as not recoverable in the form of kinetic energy from the system. It has then

$$E_a = E_r + E_d \tag{2.70}$$

The energy balance of the system, consisting of two vehicles, can then be written, for the compression phase, as

$$E = E_G + E_a \tag{2.71}$$

and then, from Eq. (2.66) we have

$$E_a = \frac{1}{2} m_c \mathbf{V}_{\mathbf{R}}^2 \tag{2.72}$$

namely, the energy absorbed during the impact depends only on the closing speed of the two vehicles at the beginning of impact.

Restitution phase

In this phase of the energy absorbed by the structures in the form of elastic potential energy, it is released, turning into kinetic energy. The extent of the release depends exclusively on the type of vehicle structures. The relative speed of vehicles, from zero at the beginning of this phase, assumes a finite value at the end of the impact.

The kinetic energy of the system at the end of the impact is

$$\overline{E} = \frac{1}{2}(m_1 + m_2)\mathbf{V}_{\mathbf{G}}^2 + \frac{1}{2} m_c \overline{\mathbf{V}}_{\mathbf{R}}^2 \tag{2.73}$$

The energy balance of the system in this second phase can then be written as

$$\overline{E} = E_G + E_r \tag{2.74}$$

The amount E_r of energy elastically restored will then

$$E_r = \frac{1}{2} m_c \overline{\mathbf{V}}_{\mathbf{R}}^2 \tag{2.75}$$

which, remembering the expression of the coefficient of restitution, can be written as

$$E_r = \frac{1}{2} m_c \mathbf{V}_{\mathbf{R}}^2 \varepsilon^2 \tag{2.76}$$

Kelvin's theorem

Obtaining E_d from Eq. (2.70), substituting the expressions of E_a and E_r obtained from Eqs. (2.72) and (2.75), and comparing with Eq. (2.8) is obtained

$$E_d = \mathbf{I}\frac{(\mathbf{V_R} - \overline{\mathbf{V}}_\mathbf{R})}{2} \quad (2.77)$$

This expression links the dissipated energy with the impulse of the contact forces and with the initial and final closing speeds. Since the latter have opposite signs, the term in brackets may also be seen as the average of the absolute values of the closing speed.

Eq. (2.75) expresses a theorem due to Kelvin, generally applicable to the case of three-degrees-of-freedom models: "The energy dissipated in an impulsive action is equal to the impulse multiplied by the average of the initial and final closing speeds, evaluated at the point of impact."

2.8.2 Normal and tangential dissipated energy

The dissipated energy can be distinguished in two contributes, E_{dn} due to the normal plastic deformation and E_{dt} due to shear deformation, tangential to the plane of impact, and the friction, with $E_d = E_{dn} + E_{dt}$

By applying the Kelvin theorem we have

$$E_{dn} = \frac{1}{2}I_n(V_{Rn} + \overline{V}_{Rn}) \quad (2.78)$$

which in the case of sliding impacts, recalling Eq. (2.30), becomes

$$E_{dn} = \frac{1}{2}\frac{(1 - \varepsilon_i^2)}{a - \mu b}V_{Rn}^2 \quad (2.79)$$

For E_{dt} term we can write

$$E_{dt} = \frac{1}{2}\mu I_n(V_{Rt} + \overline{V}_{Rt}) \quad (2.80)$$

From the first of Eqs. (2.25) and (2.29) the energy dissipated by friction and shear deformations in a sliding impact becomes

$$E_{dt} = \frac{1}{2}\mu I_n[2V_{Rt} + (b - \mu c)I_n] \quad (2.81)$$

It is noted that E_{dn} and E_{dt} depend on the coefficient of restitution and μ.

The dissipated energy globally is given by the sum of the terms in Eqs. (2.77) and (2.79). With some algebra, we obtain

$$E_d = E_{dn} + E_{dt} = \frac{1}{2}\frac{(1 + \varepsilon_i)}{(a - \mu b)}V_{Rn}^{\ 2}\left[(1 - \varepsilon_i) + 2\mu R + \mu\frac{(b - \mu c)}{(a - \mu b)}(1 + \varepsilon_i)\right] \quad (2.82)$$

with

$$R = \frac{V_{Rt}}{V_{Rn}} = \frac{(V_{2y} - \omega_2 x_2) - (V_{1y} - \omega_1 x_1)}{(V_{2x} + \omega_2 y_2) - (V_{1x} + \omega_1 y_1)} \tag{2.83}$$

This expression represents the energy dissipated in a sliding impact, in a more general condition, in which there are normal and tangential forces, translations, and rotations of the vehicles. Eq. (2.81) is quadratic in μ, and therefore, a μ_{max} value can be determined that corresponds to maximum energy dissipated.

Between two bodies that collide with an initial closing tangential speed, the higher the coefficient of friction, the higher the dissipated energy is during the impact. Increasing the value of the coefficient of friction over the maximum value μ_{max}, the friction developed causes the relative tangential motion ceases before the separation, that is, before the end of the impact, and then we have a full impact.

The value $|\mu| = \mu_{max}$ represents the minimum value of the coefficient of friction that causes the end of relative tangential motion before the separation of the bodies.

Higher values of μ do not have physical significance as only compatible with introduced energy into the system from outside. This value is

$$\mu_{max} = \frac{\left(R/(1+\varepsilon_i)\right)a + \left(1/(1+\varepsilon_i)\right)b}{c + \left(R/(1+\varepsilon_i)\right)b - \left(\varepsilon_i/(1+\varepsilon_i)\right)\left(b^2/a\right)} \tag{2.84}$$

The value $|\mu| = \mu_{max}$, therefore, represents the maximum value to be used with Eqs. (2.29) and (2.30) or (2.38) for sliding impacts, because for higher values a not realistic solution is obtained, as if it were supplied energy to the system. The tangential impulse would tend, that is, once ceased the sliding between the vehicles, to reverse the relative motion, which of course cannot be. Values higher than the maximum must be used with Eqs. (2.27) or (2.37) in the case of full impacts.

2.9 Scalar equations

For a one-dimensional collision, that is, where all the kinematic quantities are oriented along the same direction, from Eqs. (2.70), (2.75), and (2.76) we have

$$E_d = \frac{1}{2} m_c (1 - \varepsilon^2) \mathbf{V}_\mathbf{R}^\mathbf{2} \tag{2.85}$$

This expression correlates the energy dissipated to the initial closing speed of the vehicles, which generally is an unknown problem. For this purpose, in the reconstruction of road accidents, it is useful to explicit the unknown closing speed:

$$V_R = \sqrt{\frac{2E_d}{m_c(1-\varepsilon^2)}} \tag{2.86}$$

The previous expression allows calculating the closing speed at impact, knowing the deformation energy of both vehicles, for a one-dimensional collision.

The velocity change experienced by individual vehicle, substituting Eq. (2.86) into Eq. (2.10) is

$$\Delta V_1 = \frac{1}{m_1}\sqrt{2E_d m_c \frac{(1+\varepsilon)}{(1-\varepsilon)}} \tag{2.87}$$

For generic impacts, centered or oblique, with the vehicles schematized as rigid bodies in three degrees of freedom, as illustrated in Fig. 2.8. The dissipated kinetic energy associated with the translation results from the difference between initial and final translation kinetic energy:

$$E_d^T = \frac{1}{2}m_1(\mathbf{V}_1{}^2 - \overline{\mathbf{V}}_1{}^2) + \frac{1}{2}m_2(\mathbf{V}_2{}^2 - \overline{\mathbf{V}}_2{}^2) \tag{2.88}$$

Recalling

$$\begin{aligned} \Delta\mathbf{V_1} &= \overline{\mathbf{V}}_\mathbf{1} - \mathbf{V_1} \\ \Delta\mathbf{V_2} &= \overline{\mathbf{V}}_\mathbf{2} - \mathbf{V}_2 \end{aligned} \tag{2.89}$$

and considering Eq. (2.2), from Eq. (2.88) we get

$$E_d^T = \frac{1}{2}m_1\left(1+\frac{m_1}{m_2}\right)(\Delta\mathbf{V}_1)^2 + m_1\Delta\mathbf{V}_1\cdot(\overline{\mathbf{V}}_2 - \overline{\mathbf{V}}_1) \tag{2.90}$$

Also the kinetic energy of rotation dissipated in the impact results from the difference between initial and final kinetic energy:

$$E_d^R = \frac{1}{2}J_1(\omega_1{}^2 - \overline{\omega}_1{}^2) + \frac{1}{2}J_2(\omega_2{}^2 - \overline{\omega}_2{}^2) \tag{2.91}$$

As

$$\begin{aligned} \mathbf{I_1}\hat{\mathbf{h}}_\mathbf{1} &= -J_1(\overline{\omega}_1 - \omega_1) \\ \mathbf{I_2}\hat{\mathbf{h}}_\mathbf{2} &= -J_2(\overline{\omega}_2 - \omega_2) \end{aligned} \tag{2.92}$$

where $\mathbf{h}$ indicates the distance vector between the center of gravity of each vehicle and the action line of the impulse $\mathbf{I}$. Indicated with k the gyratory radius $k^2 = J/m$, Eq. (2.92) can be written as

$$\begin{aligned} (\overline{\omega}_1 - \omega_1) &= \Delta\mathbf{V}_1\hat{\mathbf{h}}_\mathbf{1}\frac{1}{k_1^2} \\ (\overline{\omega}_2 - \omega_2) &= \Delta\mathbf{V}_2\hat{\mathbf{h}}_\mathbf{2}\frac{1}{k_2^2} \end{aligned} \tag{2.93}$$

from which Eq. (2.91), similarly to what has been done for the dissipated translational energy, can be written as

$$E_d^R = \frac{1}{2} m_1 (\mathbf{\Delta V}_1)^2 \left[\left(\frac{\mathbf{h_1^2}}{k_1^2} \right) + \left(\frac{m_1}{m_2} \right) \left(\frac{\mathbf{h_2^2}}{k_2^2} \right) \right] + m_1 \mathbf{\Delta V}_1 \hat{\mathbf{h_1}} \cdot \overline{\omega}_1 + m_1 \mathbf{\Delta V}_1 \hat{\mathbf{h_2}} \cdot \overline{\omega}_2) \tag{2.94}$$

The deformation energy of the two vehicles is equal to the sum of the dissipated kinetic energy of rotation and translation:

$$E_d = E_d^T + E_d^R \tag{2.95}$$

By adding Eqs. (2.90) and (2.94), two components can be identified:

$$E_\Delta = \frac{1}{2} \left(\frac{m_1}{m_2} \right) (\mathbf{\Delta V}_1)^2 \left(\frac{m_2}{\gamma_1} + \frac{m_1}{\gamma_2} \right) \tag{2.96}$$

$$E_\varepsilon = m_1 \Delta V_1 (\overline{\mathbf{V}}_2 \cdot \mathbf{i} + h_2 \overline{\omega}_2 - \overline{\mathbf{V}}_1 \cdot \mathbf{i} + h_1 \overline{\omega}_1)$$

having indicated with $\mathbf{i}$ the unit vector in the direction of the impulse $\mathbf{I}$ and with γ the mass reduction factor given by Eq. (2.45).

By indicating with u the velocity component, in the direction of the impulse $\mathbf{I}$, of the center of impact on each vehicle we have

$$\begin{aligned} \overline{u}_1 &= \overline{\mathbf{V}}_1 \cdot \mathbf{i} - h_1 \overline{\omega}_1 \\ \overline{u}_2 &= \overline{\mathbf{V}}_2 \cdot \mathbf{i} + h_2 \overline{\omega}_2 \end{aligned} \tag{2.97}$$

If at the end of the impact, a common center of impact speed along the pulse direction between the vehicles is reached, then $\overline{u}_1 = \overline{u}_2$, and the term E_ε is zero, and the deformation energy of the two vehicles is given by $E_d = E_\Delta$, from which the vehicle speed variation module can be obtained:

$$\Delta V_1 = \frac{1}{m_1} \sqrt{2 E_d m_c^*} \tag{2.98}$$

with m_c^* given by Eq. (2.51).

The direction of the velocity variation coincides with the direction of the impulse acting on the vehicle. A similar equation can be written for vehicle 2.

If we assume that the center of impact speed at the end of the impact, calculated in the direction of the impulse, is different between the two vehicles, then we can assume a value of the restitution coefficient ε_i such that

$$\overline{u}_2 - \overline{u}_1 = \varepsilon_i (u_2 - u_1) \tag{2.99}$$

the term E_ε is no longer zero in this case, the total deformation energy, E_d, is given by the sum of both terms of Eq. (2.96), from which the speed variation can be derived as

$$\Delta V_1 = \frac{1}{m_1}\sqrt{\frac{2E_d m_2(1+\varepsilon_i)}{(1-\varepsilon_i)}} \tag{2.100}$$

This last expression is analogous to Eq. (2.87), obtained in the case of centered impact. The expression was obtained without making any assumptions about the structural characteristics of the vehicles and the modality of deformation as a function of the applied force; here, however, the mass reduction factors γ are present, as if the impulse vector does not pass through the center of gravity, a couple is produced. The couple produces a rotation, and therefore, with equal deformation, the vehicle has a smaller variation in translation speed since a part of the initial kinetic energy converts to the kinetic energy of rotation.

The deformation energy can be evaluated starting from the deformations of the vehicles, as shown later.

Comparing Eqs. (2.98) and (2.100), it results that in the case of a partially elastic impact, the expression of ΔV is multiplied by a factor f:

$$f = \sqrt{\frac{(1+\varepsilon_i)}{(1-\varepsilon_i)}} \tag{2.101}$$

which, for small values of ε_i can be developed in the binomial form:

$$f = 1 + \varepsilon_i \tag{2.102}$$

Since ε_i is nonnegative, it is clear that the inelastic impact approximation always leads to a low estimate of the vehicle speed variation. In collisions at medium-high speed the error committed is however small, given that the value of the restitution coefficient is low.

Recalling Eq. (2.7), as the direction of $\mathbf{\Delta V}$ is along the PDOF direction, by also projecting $\mathbf{V_R}$ in the direction of the PDOF, we get a scalar equation:

$$V_{R\_PDOF} = \overline{V}_{R\_PDOF} + |\Delta V_2 - \Delta V_1| \tag{2.103}$$

The coefficient of restitution, calculated with reference to the velocity of the vehicles' center of mass, is

$$\varepsilon = -\frac{\overline{V}_{R\_PDOF}}{V_{R\_PDOF}} \tag{2.104}$$

then replacing in Eq. (2.103):

$$V_{R\_PDOF} = \frac{(\Delta V_1 - \Delta V_2)}{(1+\varepsilon)} \tag{2.105}$$

As $\mathbf{\Delta V_1}$ and $\mathbf{\Delta V_2}$ have opposite sign, $|\mathbf{\Delta V_2} - \mathbf{\Delta V_1}| = \Delta V_1 + \Delta V_2$, and by replacing Eq. (2.100) by Eq. (2.105), it is possible to get the equation for the component along the PDOF direction of the relative impact velocity, valid for any impact:

$$V_{R\_PDOF} = \frac{1}{(1+\varepsilon)m_c}\sqrt{2E_d m_c^* \frac{(1+\varepsilon_i)}{(1-\varepsilon_i)}} \tag{2.106}$$

Replacing Eq. (2.50) by Eq. (2.106), Eq. (2.107) is obtained, which is a generalization of Eq. (2.86), where reduced masses of the vehicles instead of the vehicle masses appear.

$$V_{R\_PDOF} = \sqrt{\frac{2E_d}{m_c^*(1-\varepsilon_i^2)}} \tag{2.107}$$

This equation is valid for any impact, under the hypothesis, stated for Eq. (2.50), that the angular velocity of both vehicles is zero before the impact.

In each collision, for both vehicles, the three vectors, initial velocity, final and the change of speed, form a triangle, as shown in Fig. 2.15.

Eq. (2.87) can be used in the triangle of speed to find the speed unknowns, knowing the direction of the $\mathbf{\Delta V}$.

2.10 Speed change in the center of the impact

Referring to Fig. 2.16, we have

$$\mathbf{\Delta V}' = \mathbf{\Delta V} + \Delta\omega\,\hat{\mathbf{d}} \tag{2.108}$$

where $\mathbf{d}$ is the vector applied in the center of gravity, which connects the center of impact.

Recalling Eq. (2.93), we have

$$\mathbf{\Delta V}' = \mathbf{\Delta V} + \frac{1}{k^2}(\mathbf{\Delta}\hat{\mathbf{V}}\hat{\mathbf{h}}\mathbf{d}) \tag{2.109}$$

which provides the relationship between the velocity variation of the center of gravity of the vehicle and the velocity variation of the center of impact.

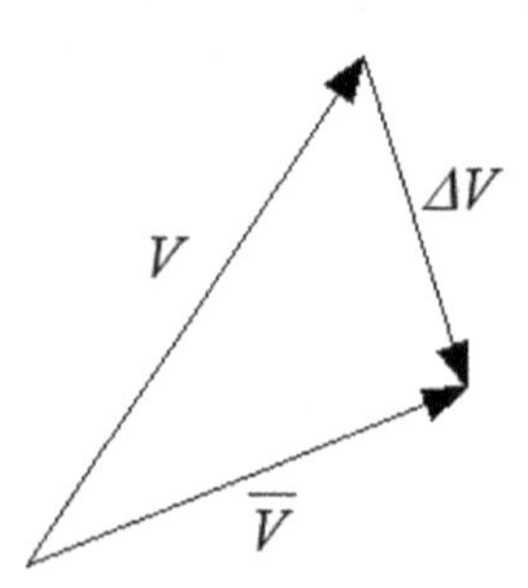

FIGURE 2.15 The triangle of speeds.

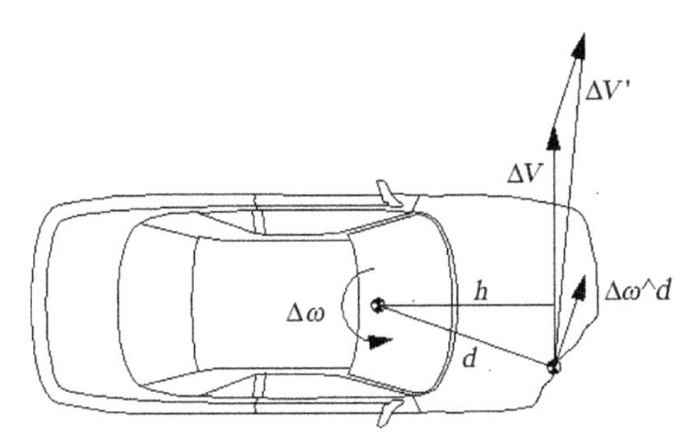

FIGURE 2.16 Polygon of speed in the center of impact.

As can be seen, the two vectors are not generally aligned on the same straight line.

The component of the speed variation in the center of impact, $\mathbf{\Delta V'_I}$, in the impulse direction, is

$$\mathbf{\Delta V'_I} = \mathbf{\Delta V} + \mathbf{\Delta \omega}\,\hat{\mathbf{h}} \tag{2.110}$$

From Eq. (2.93), in modulus, we have

$$\Delta V'_I = \Delta V + \Delta V \frac{h^2}{k^2} \tag{2.111}$$

or

$$\Delta V = \gamma\, \Delta V'_I \tag{2.112}$$

which gives the relationship between the speed change of the center of gravity of the vehicle and that of the center of impact, in the pulse direction. From Eq. (2.2), we obtain

$$\frac{\Delta V'_{I1}}{\Delta V'_{I2}} = -\frac{m_2 \gamma_2}{m_1 \gamma_1} \tag{2.113}$$

2.11 Constant acceleration circle

As the impact duration is unique for all the points of the vehicle, then, from Eq. (2.112), we have

$$\frac{\Delta V}{\Delta t} = \gamma \frac{\Delta V'_I}{\Delta t} \tag{2.114}$$

namely,

$$a_G = \gamma a'_I \tag{2.115}$$

The discussion can be generalized extending the analysis to a generic point Q, anywhere in the vehicle.

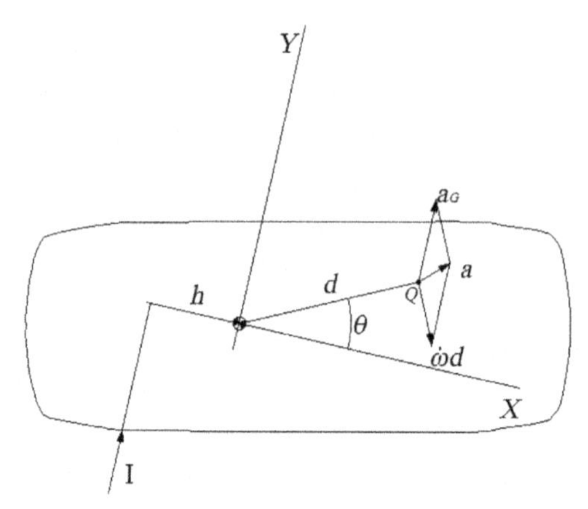

FIGURE 2.17 Acceleration components at point Q.

Referring to Fig. 2.17, in which the $X-Y$ reference system, centered in the center of gravity, has been chosen, so that the Y-axis is aligned with the direction of the applied pulse, the acceleration of the point Q (x, y) is

$$\boldsymbol{a} = \boldsymbol{a}_G + \dot{\omega}\,\hat{\mathbf{d}} \tag{2.116}$$

The components of this acceleration along the axes X and Y is

$$\begin{aligned} a_X &= d\dot{\omega}\sin(\theta) \\ a_Y &= a_G - d\dot{\omega}\cos(\theta) \end{aligned} \tag{2.117}$$

while the module is

$$a^2 = a_X^2 + a_Y^2 = \left[a_G^2 + d^2\dot{\omega}^2 - 2a_G d\dot{\omega}\cos(\theta)\right] \tag{2.118}$$

Since $Ih = J\Delta\omega$ and $I = F\Delta t$, remembering that $J = mk^2$ we have

$$\dot{\omega} = \frac{Fh}{mk^2} = a_G\frac{h}{k^2} \tag{2.119}$$

Substituting Eq. (2.119) into Eq. (2.118), we have

$$a^2 = a_G^2\left[1 + d^2\left(\frac{h}{k^2}\right)^2 - 2d\frac{h}{k^2}\cos(\theta)\right] \tag{2.120}$$

We define c amplification factor of the acceleration quantity:

$$c = \frac{a}{a_G} \tag{2.121}$$

As $d^2 = x^2 + y^2$ and $d\cos(\theta) = x$, Eq. (2.120) becomes

$$c^2 = \left[1 + (x^2 + y^2)\left(\frac{h}{k^2}\right)^2 - 2x\frac{h}{k^2}\right] \tag{2.122}$$

multiplying both members of Eq. (2.122) for $\left(k^2/h\right)^2$, with some step we have

$$\left(x-\frac{k^2}{h}\right)^2+y^2=\left(c\frac{k^2}{h}\right)^2 \tag{2.123}$$

which is the equation of a circle with center in the point P of coordinates $\left(k^2/h,0\right)$ and radius:

$$\rho=c\frac{k^2}{h} \tag{2.124}$$

So, when an impulse $\mathbf{I}$ is applied to the vehicle, all the points that are on that circumference undergo the same acceleration.

The center of the circle, that is, for $\rho=0$, has a zero acceleration, since from Eq. (2.124), we have $c=0$, while all the other points of the vehicle undergo an acceleration proportional to the distance from the center of the circumference. At each point the acceleration direction is tangent to the circle passing from the point.

The acceleration amplification factor c for a given point of coordinates (x, y), may be derived from Eq. (2.124) by observing that, as shown in Fig. 2.18:

$$\rho=\sqrt{(x_P-x)^2+y^2} \tag{2.125}$$

The determination of the circles of constant acceleration allows identifying the most critical areas of the vehicle during the impact, for example, concerning the stress suffered by the occupants or the location of the sensors for the passive safety systems activation such as the airbag sensors.

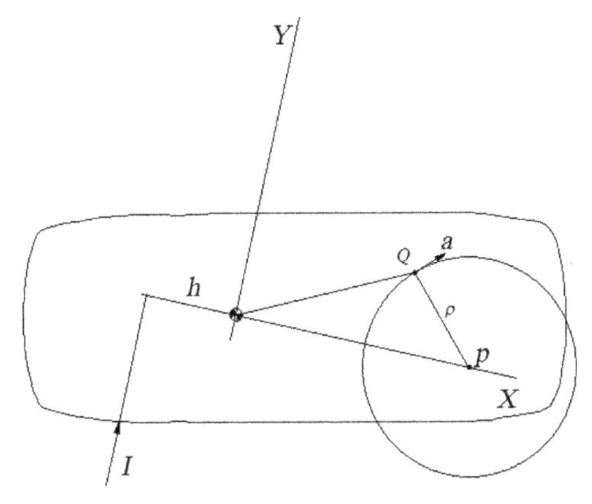

FIGURE 2.18 Circle of constant acceleration passing through the point Q.

References

Amato, G., O'Brien, F., Ghosh, B., Simms, C.K., 2013. Multibody modelling of a TB31 and a TB32 crash test with vertical portable concrete barriers: model verification and sensitivity analysis. Proc. Inst. Mech. Eng., K: J. Multi-body Dyn. 227 (3), 245–260.

Antonetti, V.W., 1998. Estimating the Coefficient of Restitution of Vehicle-to-Vehicle Bumper; Impacts Amatech Review; SAE Paper 980552.

Brach, R.M., 1983. Analysis of planar vehicle collisions using equations of impulse and momentum. Accid. Anal. Prev. 15 (2), 105–120.

Brach, R., Brach, R., 2005. Vehicle Accident and Reconstruction Methods. Society of Automotive Engineers -SAE International, Warrendale, PA. ISBN 0-7680-0776-3.

Chen, D.Y., Wang, L.M., Wang, C.Z., Yuan, L.K., Zhang, T.Y., Zhang, Z.Z., 2015. Finite element based improvement of a light truck design to optimize crashworthiness. Int. J. Automot. Technol. 16 (1), 39–49.

El Kady, M., Elmarakbi, A., MacIntyre, J., Alhari, M., 2016. Multi-body integrated vehicle-occupant models for collision mitigation and vehicle safety using dynamics control systems. Int. J. Syst. Dyn. Appl. 5 (2), 80–122.

Gilardi, G., Sharf, I., 2002. Literature survey of contact dynamics modelling. Mech. Mach. Theory 37, 1213–1239.

Goldsmith, W., 2001. Impact—The Theory and Physical Behaviour of Colliding Solids. Dover Publications, Inc., Mineola, NY.

Hamza, K., Saitou, K., 2003. Design optimization of vehicle structures for crashworthiness using equivalent mechanism approximations. In: ASME 2003 International Design Engineering Technical Conferences and Computers and Information in Engineering Conference, no. DETC2003/DAC-48751, American Society of Mechanical Engineers, Chicago, IL, pp. 459–472.

Han, I., 2015. Impulse-momentum based analysis of vehicle collision accidents using Monte Carlo simulation methods. Int. J. Automot. Technol. 16 (2), 253–270.

Huang, M., 2002. Vehicle Crash Mechanics. CRC Press, Boca Raton, FL.

Iraeus, J., Lindquist, M., 2015. Pulse shape analysis and data reduction of real-life frontal crashes with modern passenger cars. Int. J. Crashworthiness 20 (6), 535–546.

Ishlkawa, H., 1993. Impact Model for Accident Reconstruction—Normal and Tangential Restitution Coefficients, SAE Paper 930654.

Ishikawa, H., 1994. Energy loss and delta-V in vehicle collision: car-to-car side impact. JSAE Rev. 15 (3), 215–221.

Jonsén, P., Isaksson, E., Sundin, K., Oldenburg, M., 2009. Identification of lumped parameter automotive crash models for bumper system development. Int. J. Crashworthiness 14 (6), 533–541.

Klausen, A., Tørdal, S.S., Karimi, H.R., Robbersmyr, K.G., Jecmenica, M., Melteig, O., 2014. Firefly optimization and mathematical modeling of a vehicle crash test based on single-mass. J. Appl. Math. 2014, 1–10. Available from: https://doi.org/10.1155/2014/150319.

Kolk, H., Tomasch, E., Sinz, W., Bakker, J., Dobberstein, J., 2016. Evaluation of a momentum based impact model and application in an effectivity study considering junction accidents. In: Proceedings of International Conference "ESAR—Expert Symposium on Accident Research", Hannover, Germany, June 9, 2016.

Kudlich, H., 1966, Beitrag zur Mechanik des Kraftfahrzeug-Verkehrsunfalls. Techn. Hochsch., Diss.-Wien.

Marquard, E., 1962. On the mechanics of vehicle collisions. Automobiltechnische Zeitschrift 64 (5), 142–148.

Marzbanrad, J., Pahlavani, M., 2011. Calculation of vehicle-lumped model parameters considering occupant deceleration in frontal crash. Int. J. Crashwothiness 16 (4), 439–455.

Ofochebe, S.M., Ozoegwu, C.G., Enibe, S.O., 2015. Performance evaluation of vehicle front structure in crash energy management using lumped mass spring system. Adv. Model. Simul. Eng. 2 (2), 1–18.

Pahlavani, M., Marzbanrad, J., 2015. Crashworthiness study of a full vehicle-lumped model using parameters optimisation. Int. J. Crashworthiness 20 (6), 573–591.

Pawlus, W., Karimi, H.R., Robbersmyr, K.G., 2011a. Application of viscoelastic hybrid models to vehicle crash simulation. Int. J. Crashworthiness 16 (2), 195–205.

Pawlus, W., Karimi, H.R., Robbersmyr, K.G., 2011b. Development of lumped-parameter mathematical models for a vehicle localized impact. J. Mech. Sci. Technol. 25 (7), 1737–1747.

Sousa, L., Verssimo, P., Ambrsio, J., 2008. Development of generic multibody road vehicle models for crashworthiness. Multibody Syst. Dyn. 19, 133–158.

Vangi, D., 2008. Ricostruzione della dinamica degli incidenti stradali—principi e applicazioni. Firenze University Press, p. 380. ISBN: 978-88-8453-783-6.

Vangi, D., Mastandrea, M., 2005. Influence of braking force in low speed vehicle collisions. Proc. Inst. Mech. Eng., D: J. Automob. Eng. 219 (2), 151–164.

Varat, M., Husher, S.E., 2003. Crash pulse modelling for vehicle safety research. In: 18th Int. Tech. Conf. Enhanced Safety Vehicles (ESV), no. 501, National Highway Traffic Safety Administration.

Yildiz, A.R., Solanki, K.N., 2012. Multi-objective optimization of vehicle crashworthiness using a new particle swarm based approach. Int. J. Adv. Manuf. Technol. 59 (1–4), 367–376.

Further reading

Vangi, D., 2009. Energy loss in vehicle to vehicle oblique impact. Int. J. Impact Eng. 36 (3), 512–521. ISSN 0734-743X.

Chapter 3

Models for the structural vehicle behavior

Chapter Outline

3.1 Lumped mass models

The lumped mass models describe the impact behavior by schematizing the vehicle with one or more rigid masses, connected by load paths consisting of zero mass elements. For a correct description of the behavior of the vehicle during the collision, it is necessary to extract optimal lumped parameters from actual or simulated vehicle crash event data. Modeling a vehicle, using a lumped parameter method, allows deriving the differential equations that describe the dynamic behavior of the system using Newton's second law and, in particular, to obtain the $F(x)$ curves that express the relationship between force and deformation. These curves are related to the vehicle's structural response, for a given area affected by the impact, and can be considered characteristics of each vehicle or structurally similar vehicle classes.

3.1.1 Mass–spring model (Campbell model)

The most straightforward function for approximating the curve $F(x)$ is the straight line, as a pure mass–linear spring model. Many accident reconstruction software, widely used today as the CRASH4-like software, employ this type of schematization to describe the relationship between force and deformation of a vehicle during the compression phase (see Fig. 3.1).

Vehicle Collision Dynamics. DOI: https://doi.org/10.1016/B978-0-12-812750-6.00003-2

Usually, the force is normalized with respect to the width L of the deformed area, which in the case of collisions against the barrier coincides with the width of the front.

Fig. 3.2 shows the linear relationship between normalized force and deformation, relative only to the compression phase. The deformation is distinct in permanent deformation C (crush) and elastic deformation C_e.

A minimum value of force, indicated by the constant A, is assumed, below which there are no permanent deformations. For higher contact forces the force increases with a positive slope, indicated by the constant B, and the linear expression of the normalized force as a function of the residual deformation is

$$F = BC + A \tag{3.1}$$

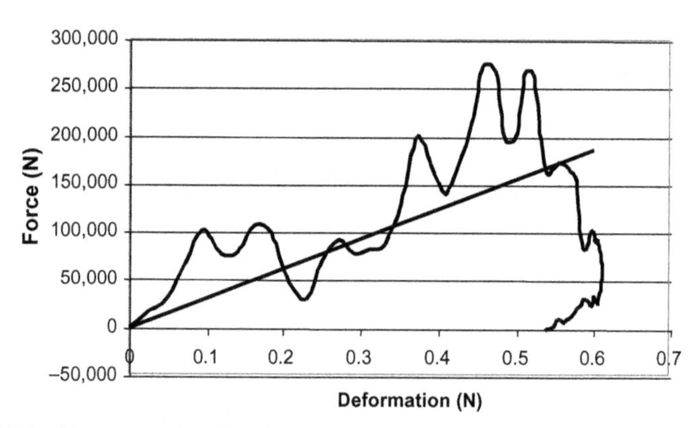

FIGURE 3.1 Linear approximation of the force−deformation curve.

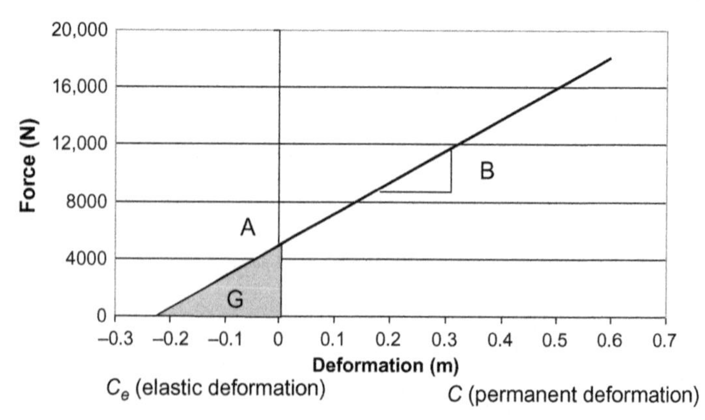

FIGURE 3.2 Normalized force versus residual deformation curve.

The area under the curve, in part relating to elastic deformations, is indicated with G and represents the normalized energy absorbed elastically before the occurrence of permanent deformations and is equal to

$$G = \frac{A^2}{2B} \tag{3.2}$$

The constants A and B, as well as the $F(C)$ curve, are characteristics for each vehicle and a given area of impact and can be determined through regression of many experimental results obtained from crash tests at different speeds against the barrier.

Vehicle-to-barrier impact

The linear pattern of the force as a function of the deformation is also congruent with the linear trend between impact speed and residual deformation, experimentally observed first by Campbell based on the results of numerous tests against the crash barrier (Campbell, 1974). The impact phase of the vehicle against a fixed barrier can be schematized, in fact, with a mass–spring model with 1 degree of freedom (see Fig. 3.3), in which the spring has stiffness $k = BL$; the spring has a plastic behavior, that is, the energy is not accumulated in the form of elastic potential energy but is dissipated in deformation; it is considered, therefore, only the compression phase. We can write the dynamic equilibrium equation as

$$m\ddot{x} + kx = 0 \tag{3.3}$$

The general solution of Eq. (3.3) is

$$x(t) = a_1 \sin \omega t + a_2 \cos \omega t \tag{3.4}$$

were

$$\omega = \sqrt{\frac{k}{m}} \tag{3.5}$$

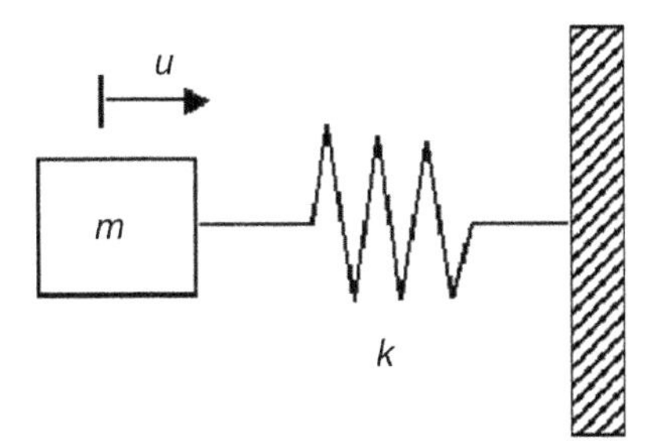

FIGURE 3.3 Model to a degree of freedom bump against the barrier.

With the following initial conditions:

$$\begin{aligned} x(0) &= 0 \\ \dot{x}(0) &= V \end{aligned} \tag{3.6}$$

we get

$$\begin{aligned} x &= \frac{V}{\omega}\sin \omega t \\ \dot{x} &= V \cos \omega t \\ \ddot{x} &= -\omega V \sin \omega t \end{aligned} \tag{3.7}$$

The mass–spring model gives a sine pulse acceleration response (a quarter sine pulse). The maximum deformation of the vehicle corresponds to the maximum displacement X of the vehicle against the barrier and is given by

$$X = \frac{V}{\omega} \tag{3.8}$$

that can be written as

$$V = \omega X \tag{3.9}$$

Eq. (3.9) compared with Eq. (1.9) indicates that the reciprocal of ω coincides with the centroid time t_C:

$$t_C = \frac{1}{\omega} \tag{3.10}$$

Eq. (3.9) indicates the searched linear relationship between the initial speed and maximum deformation. Also in this case one can assume a minimum speed value, below which there is no permanent deformation, obtaining the aforementioned empirical relationship (see Fig. 3.4) between impact

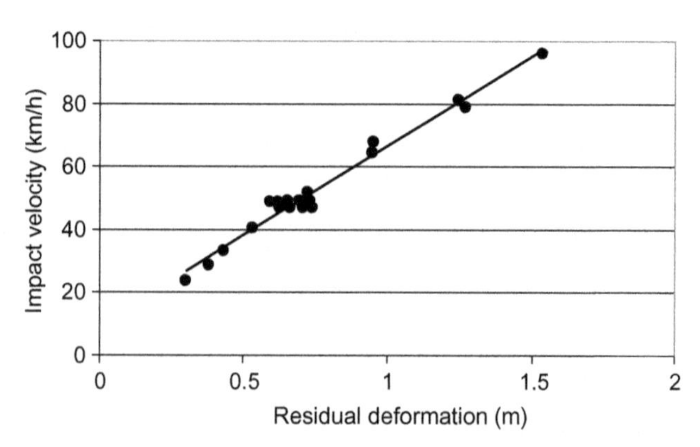

FIGURE 3.4 Experimental trend observed between impact speed and permanent deformation.

velocity and the permanent deformation of the vehicles C, expressed as follows:

$$V = b_1 C + b_0 \tag{3.11}$$

where

$$\omega = b_1 \tag{3.12}$$

About the force per unit of width of the deformed area, we also get

$$F = \frac{m}{L} a = \frac{m}{L} b_1 V \tag{3.13}$$

It should be noted that the width L is the one referred to as the stiffness k in Eq. (3.5). For example, in a frontal collision, L can be the width of the entire front or of the frontal part between the edges of the bumper, identified through a straight line at 45 degrees, as shown in Fig. 3.5. For side impact, L can be assumed equal to the length of the vehicle, always referring to the length between the edges of the bumper, identified through a straight line at 45 degrees.

From Eq. (3.11), Eq. (3.13) becomes

$$F = \frac{m}{L} \left(b_1 b_0 + b_1^2 C \right) \tag{3.14}$$

From the latter, compared with Eq. (3.1), we obtain the following relations between the various coefficients:

$$A = \frac{m}{L} b_0 b_1 \quad B = \frac{m}{L} {b_1}^2 \tag{3.15}$$

From the Eqs. (3.10), (3.12), and the second of (3.15), it is evident that the centroid time, that is, the time corresponding to the centroid of the area under the acceleration curve from zero to the moment of maximum dynamic

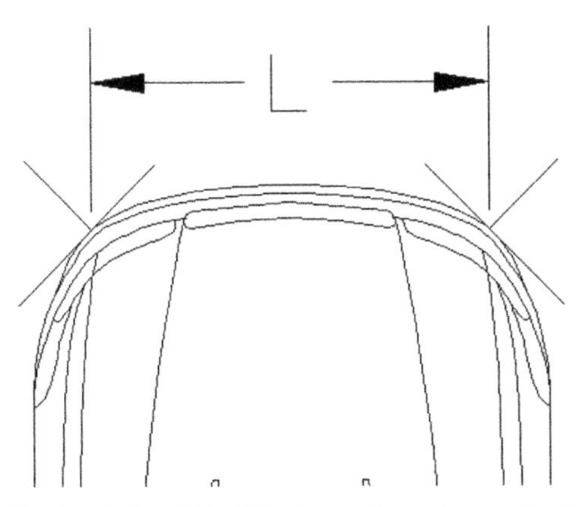

FIGURE 3.5 Identification of the width of the damaged area due to a frontal impact.

deformation is linked to the stiffness of the vehicle. The higher the stiffness of the vehicle, the smaller the centroid time is.

The mass–spring model is widely used in the analysis and reconstruction of road accidents and in particular for the evaluation of kinetic energy dissipated by vehicle deformations.

Since the coefficients A and B are related to the unit of width of the deformed area, the overall kinetic energy absorbed during deformation is equal to

$$E_a = \int_0^L \left[G + \int_0^C F(C)dC \right] dl = \int_0^L \left(G + AC + \frac{BC^2}{2} \right) dl \qquad (3.16)$$

By neglecting the elastic restitution, $E_d = E_a$ and from Eqs. (2.100) and (2.107), it is possible to calculate the closing speed and the speed variations of the vehicles.

Vehicle-to-vehicle impact, effective mass system

Let us assume we have two vehicles, each of which can be schematized as a mass m_i $(i = 1, 2)$ and a spring of stiffness k_i. In the case of vehicle–vehicle impact the system can be schematized as shown in Fig. 3.6.

$V_R = V_1 - V_2$ gives the relative speed of the approach of the vehicles, and the relative movement is given by $\alpha\ (t) = x_1(t) - x_2(t)$.

The system of two masses can be studied considering an equivalent system, called an effective mass system, schematized in Fig. 3.7, as if it were a single vehicle against a fixed barrier.

The effective mass system can be studied with the equations seen in the previous paragraph, relating to the impact of the vehicle against a fixed barrier, using an effective mass of

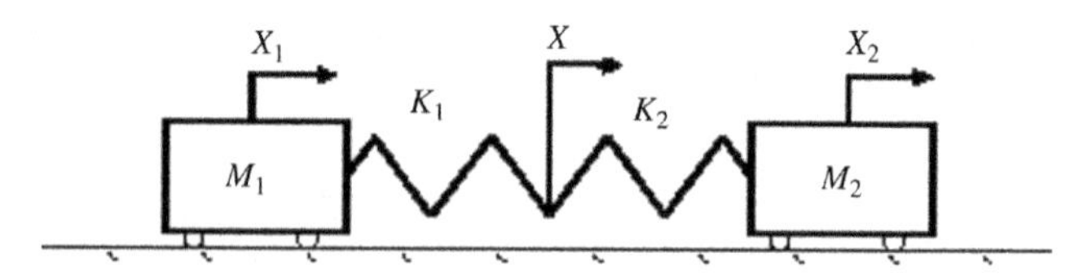

FIGURE 3.6 A two mass system.

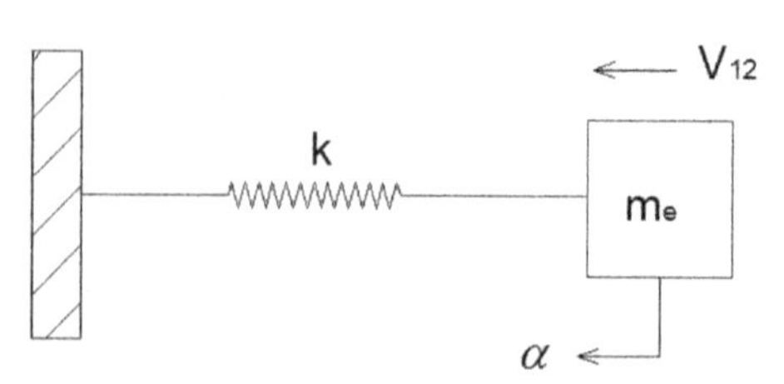

FIGURE 3.7 Effective mass system.

$$m_c = \frac{m_1 m_2}{m_1 + m_2} \tag{3.17}$$

The stiffness k of the spring is equal to the stiffness of the two springs in series:

$$k_c = \frac{k_1 k_2}{k_1 + k_2} \tag{3.18}$$

while the displacement is equal to $\alpha(t) = x_1(t) - x_2(t)$ and the speed is equal to the relative speed $V_R = V_1 - V_2$.

The equation of motion of the system can be written as

$$m_c \ddot{\alpha} + k_c \alpha = 0 \tag{3.19}$$

The relative acceleration, speed, and displacement of the system are

$$\begin{aligned} \alpha(t) &= \frac{V}{\omega_c} \sin(\omega_c t) \\ \dot{\alpha}(t) &= V \cos(\omega_c t) \\ \ddot{\alpha}(t) &= -\omega_c V \sin(\omega_c t) \end{aligned} \tag{3.20}$$

where angular natural frequency results

$$\omega_c = \sqrt{\frac{k_c}{m_c}} \tag{3.21}$$

The absolute accelerations of the two vehicles are

$$\begin{aligned} \ddot{x}_1(t) &= g_1 \ddot{\alpha}(t) \\ \ddot{x}_2(t) &= g_2 \ddot{\alpha}(t) \end{aligned} \tag{3.22}$$

and similarly for speeds and displacements, where the terms g represent the mass reduction factors:

$$\begin{aligned} g_1 &= \frac{m_2}{m_1 + m_2} \\ g_2 &= \frac{m_1}{m_1 + m_2} \end{aligned} \tag{3.23}$$

If the mass m_2 becomes infinite (fixed barrier) the actual mass is equal to m_1, g_1 becomes equal to 1 and g_2 equal to 0 and the system is equal to that seen in the previous paragraph.

The duration of the collision between two vehicles depends by the stiffness and mass of the vehicles themselves. The two vehicles at the time of impact have a speed of opposite sign, that is, a closing speed that tends to decrease the distance between the centers of gravity of the vehicles by deforming the structures. At the time of maximum dynamic deformation the vehicles reach a common velocity, that is, their relative speed became 0.

The instant at which this occurs can be calculated by putting equal to 0 the second of Eq. (3.20), obtaining

$$t_X = \frac{\pi}{2}\sqrt{\frac{m_c}{k_c}} \tag{3.24}$$

or considering the linearity between force and deformation and the fact that in every moment the forces acting on the vehicle must be equal to each other:

$$t_X = \frac{\pi}{2}\sqrt{m_C\frac{(X_1+X_2)}{X_1X_2}} \tag{3.25}$$

where t_X is the duration of the compression phase of impact between two vehicles (time at maximum dynamic deformation).

In the case of oblique impact, in Eqs. (3.24) and (3.25), $m_c{}^*$ is put in place of m_c.

If we put m_2 and k_2 equal to ∞, we obtain the compression phase duration of an impact against the barrier, also referred to as BET (barrier equivalent time):

$$\text{BET} = t_{Xb} = \frac{\pi}{2}\sqrt{\frac{m_1}{k_1}} \tag{3.26}$$

Recalling Eqs. (3.5) and (3.10), the BET is related to the centroid time t_c as

$$t_{Xb} = \frac{\pi}{2}t_c = 0.64\,t_c \tag{3.27}$$

From Eqs. (3.24) and (3.25), it can be observed that the impact duration (or better, of the compression phase) may be different from the "equivalent" duration against the barrier, that is, from the duration of an impact against the barrier of one of two vehicles.

For example, in a collision between two vehicles with the same mass and stiffness, the impact duration is equal to that which would result in a collision with the barrier, for the same extent of the deformation. In general, if vehicle 2 has a mass less than that of vehicle 1, the duration of the impact between the two vehicles decreases compared to that of vehicle 1 against the barrier and vice versa. In the opposite trend, if vehicle 2 has stiffness lower than vehicle 1, then the duration of impact between the two vehicles increases compared to that of vehicle 1 against the barrier.

Coefficient of restitution in the collision between two vehicles

In a collision between two vehicles, during the elastic restitution phase, the energy recovered is linked to that absorbed by

$$E_r = E_a \varepsilon^2 \tag{3.28}$$

This is the energy recovered globally due to the elastic restitution of the structures of both vehicles; it can then write that the energy globally recovered is equal to the sum of the energies recovered by each vehicle:

$$E_r = E_{r1} + E_{r2} = E_{a1}\varepsilon_1^2 + E_{a2}\varepsilon_2^2 \tag{3.29}$$

where ε_1 indicates the coefficient of restitution of the vehicle 1, corresponding to a deformation level equal to the EES_1 (energy equivalent speed) of the vehicle and energy absorbed E_{a1}, similarly for the vehicle 2. From Eq. (3.29), we can derive the coefficient of restitution as a function of the individual values for each vehicle:

$$\varepsilon = \sqrt{\frac{E_{a1}}{E_a}\varepsilon_1^2 + \frac{E_{a2}}{E_a}\varepsilon_2^2} \tag{3.30}$$

assuming, for both vehicles a linear relationship between force F and deformation, with constant stiffness k, we can write

$$E_a = E_{a1} + E_{a2} = \frac{1}{2}\frac{F^2}{K_1} + \frac{1}{2}F^2\left(\frac{k_1 + k_2}{k_1 k_2}\right) \tag{3.31}$$

and, substituting in Eq. (3.30), we write

$$\varepsilon = \sqrt{\frac{k_2}{k_1 + k_2}\varepsilon_1^2 + \frac{k_1}{k_1 + k_2}\varepsilon_2^2} \tag{3.32}$$

that provides a relationship between the stiffness of each vehicle and the overall elastic restitution. Recalling Eqs. (3.5) and (3.12), Eq. (3.32) can be written as a function of the masses and stiffness coefficients b_1 usually available in the literature:

$$\varepsilon = \sqrt{\frac{\left(m_2 b_{12}^2\right)\varepsilon_1^2 + \left(m_1 b_{11}^2\right)\varepsilon_2^2}{m_1 b_{11}^2 + m_2 b_{12}^2}} \tag{3.33}$$

that allows calculating the coefficient of restitution in a collision between two vehicles starting from the coefficients of restitution obtained in tests against a barrier for each vehicle. The coefficients of restitution of each vehicle can also be derived from ε curves as a function of impact speed [equivalent test speed (ETS)] against the barrier. These curves can be obtained by interpolating the data from tests on wheelbase classes of vehicles; in this case, knowing the EES value for the vehicle, the impact speed, which is necessary to obtain the coefficient of restitution by the abovementioned curves, must be chosen for attempts, recalling that from Eq. (4.44), we have

$$\mathrm{ETS} = \frac{\mathrm{EES}}{\sqrt{1-\varepsilon^2}} \tag{3.34}$$

3.1.2 McHenry model

The above linear schematization considers the compression phase alone, without the phase of restitution, which, however, for low-speed impacts is not negligible. Neglecting the restitution, vehicle speed variation ΔV lower than the reality is obtained. To correct this tendency to underestimate the ΔV of CRASH4-like software, McHenry (McHenry and McHenry, 1997) proposed an approximation of the $F(x)$ curve, as shown in Fig. 3.8, which allows a better estimate of the dissipated energy and the vehicle speed variations.

Linear segments characterize the curve with slopes k_1 and k_2 distinct (see Fig. 3.9); k_1 represents the stiffness of the vehicle during the compression phase, at the end of which it reaches the maximum dynamic deformation X

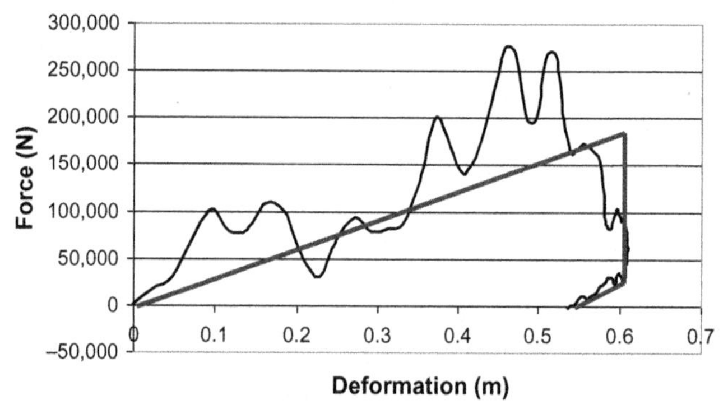

FIGURE 3.8 Linear approximation of the $F(x)$ curve, with elastic restitution.

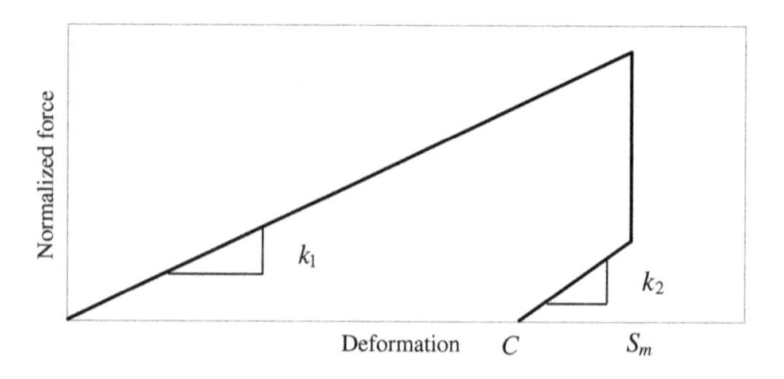

FIGURE 3.9 Coefficients of the $F(x)$ curve.

and the maximum force. At the end of this phase the force decreases abruptly until the structures begin to recover a part of the deformation, following a curve with a stiffness k_2. At the end of the restitution phase, it has a permanent deformation C.

Experimentally, the restitution coefficient can be correlated to the maximum deformation with functions of the $1/X$ or $1/X^2$ type.

In this model the coefficient of restitution ε is related to the maximum dynamic deformation through the relationship:

$$\varepsilon = \frac{\Gamma}{X} + \rho \tag{3.35}$$

where Γ and ρ are two constants. As for X, which tends to zero the value of ε tends to infinity, the curve is valid for a value less than 1. The constants are determined from the experimental data, and by imposing the condition given by Eq. (3.28), which, by observing Fig. 3.8, can be written as

$$\varepsilon = \sqrt{\frac{E_r}{E_a}} = \sqrt{\frac{k_2(X-C)^2}{k_1X^2}} = \left(1 - \frac{C}{X}\right)\sqrt{\frac{k_2}{k_1}} \tag{3.36}$$

The four coefficients k_1, k_2, Γ, and ρ can be determined starting from the coefficients A and B determined in Campbell's linear model.

A limitation of this schematization is that it requires experimental data, with also the restitution phase, which are often not available; in particular, it requires the results obtained from a series of crashes of the same vehicle, at different speeds, with an indication of the elastic restitution value obtained. The following are reported a procedure for obtaining the values of the coefficients in the case where the experimental results are available and an approximate procedure, usable if they are not available.

Procedure from the experimental data:

1. The coefficients A and B starting from the tabulated data for each wheelbase class (Siddal, and Day, 1996; Neptune, 1998) of the vehicle are determined for the given vehicle.
2. Determination of Γ. Plotting the values of the coefficient of restitution obtained from experiments for different residual deformation values, the value γ of the coefficient corresponding to a zero deformation is extrapolated. From Eq. (3.36), for $C = 0$ is obtained:

$$\varepsilon_{\max} = \sqrt{\frac{k_2}{k_1}} = \gamma \tag{3.37}$$

from which

$$k_2 = \gamma^2 k_1 \tag{3.38}$$

By requiring that the variation of the vehicle speed during the compression phase, calculated with the present model, is equal to that calculated by the Campbell's one, with a few steps [for details reference should be made to the original work of McHenry (McHenry and McHenry, 1997)], we get

$$\Gamma = \frac{A}{B}\sqrt{\frac{k_2}{k_1}} \tag{3.39}$$

that, recalling Eq. (3.37), provides the searched value as a function of the coefficients A and B:

$$\Gamma = \frac{A}{B}\gamma \tag{3.40}$$

3. Determination of k_1. By requiring that the energy absorbed in the compression phase in the present model is equal to that calculated by the Campbell model, namely, that the area under the $F(x)$ curve are equal, we have

$$E_a = \frac{1}{2}k_1X^2 = AC + \frac{BC^2}{2} + \frac{A^2}{2B} \tag{3.41}$$

from which a linear relationship between the residual deformation C and the maximum deformation X is obtained, in the form of

$$C = X\sqrt{\frac{k_1}{B}} - \frac{A}{B} \tag{3.42}$$

Laying on a graph the experimental data C in function of X and interpolating it with a straight line with an intercept equal to $-A/B$, as shown in Fig. 3.10, the angular coefficient of the straight line β is evaluated, which for Eq. (3.42) must be

$$\beta = \sqrt{\frac{k_1}{B}} \tag{3.43}$$

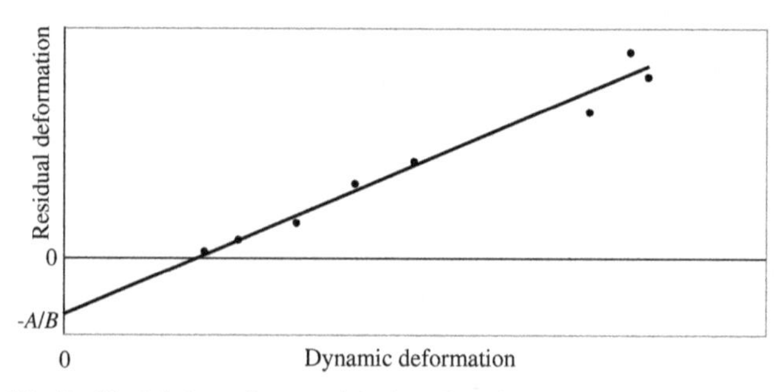

FIGURE 3.10 Residual deformation as a function of maximum deformation.

that yields the value of k_1:

$$k_1 = \beta^2 B \tag{3.44}$$

4. Determination of ρ. The dynamic deformation X_0 corresponding to the beginning of permanent deformation can be derived from Eq. (3.42) putting $C = 0$:

$$X_0 = \frac{A}{\sqrt{Bk_1}} \tag{3.45}$$

This value of the dynamic deformation is associated with the constant Γ and ρ following Eq. (3.35):

$$X_0 = \frac{\Gamma}{\gamma - \rho} \tag{3.46}$$

from which, by equating Eq. (3.46) with Eq. (3.45), and resolving in ρ, we get

$$\rho = (1 - \beta)\gamma \tag{3.47}$$

5. Determination of k_2. From Eqs. (3.38) and (3.44), we have

$$k_2 = \gamma^2 \beta^2 B \tag{3.48}$$

6. The final step comprises checking the values found, plotting ε in function of C, and using Eqs. (3.35) and (3.42). The curve thus obtained should well approximate the curve obtained in step 2.

Procedure usable in the absence of experimental data:

Very often, the experimental data to be able to draw curves referred to in steps 2 and 3, respectively, are not available. In this case, one can pursue the same steps listed earlier, with the following exceptions:

- In step 2, it is estimated the value γ, coefficient of restitution for very low-speed impacts, based on experience or on the numerous correlations reported in the literature, choosing the vehicles the most similar as possible to that under investigation; typical values are between 0.6 and 0.8.
- In step 3, the β value is estimated based on experience; in general, a unitary β value is a good approximation.

Example: Consider a vehicle subject to a crash test against the rigid barrier, with a residual deformation $C = 0.8$ m. Such a vehicle, according to its wheelbase equal to 2.61 m, falls into Class 3 of NHTSA classification (www.nhtsa.com), with coefficients

$$\mathrm{A} = 5653\ \mathrm{kg}/m$$

$$B = 39,320\ \mathrm{kg}/m^2$$

From Eq. (3.2), we get $G = 408$ kg.
As there is no experimental data, we put $\gamma = 0.7$ and $\beta = 1$.
From the above equations, we obtain parameters of the $F(x)$ curve

$$k_1 = \beta^2 B = 39,320$$

$$\Gamma = \frac{A}{B}\gamma = \frac{5653}{39,320}0.7 = 0.1$$

$$\rho = (1-\beta)\gamma = 0$$

$$k_2 = \gamma^2\beta^2 B = 0.7^2 \cdot 39,320 = 19,266$$

From the residual deformation of 0.8 m, we can calculate the coefficient of restitution and the maximum dynamic deformation that has occurred in the collision:

$$X = \left(C + \frac{A}{B}\right)\sqrt{\frac{B}{k_1}} = 0.94$$

$$\varepsilon = \frac{\Gamma}{X} + \rho = 0.106$$

As per the verification of the calculations, we can plot the ε varying the residual deformation:

A	5653
B	39,320
γ	0.7
β	1
Γ	0.100638
k_1	39,320
ρ	0
k_2	19,266.8

C	X	ε
0.00	0.14	0.70
0.05	0.19	0.52
0.10	0.24	0.41
0.15	0.29	0.34
0.20	0.34	0.29
0.25	0.39	0.26
0.30	0.44	0.23
0.35	0.49	0.20
0.40	0.54	0.19
0.45	0.59	0.17
0.50	0.64	0.16
0.55	0.69	0.15

(Continued)

(Continued)

C	X	ε
0.60	0.74	0.14
0.65	0.79	0.13
0.70	0.84	0.12
0.75	0.89	0.11
0.80	0.94	0.11
0.85	0.99	0.10
0.90	1.04	0.10
0.95	1.09	0.09
1.00	1.14	0.09
1.05	1.19	0.08
1.10	1.24	0.08
1.15	1.29	0.08
1.2	1.34	0.07
1.25	1.39	0.07
1.3	1.44	0.07
1.35	1.49	0.07
1.4	1.54	0.07
1.45	1.59	0.06

It is noted that the obtained coefficient of restitution trend is plausible and comparable with the common trend that is found in the literature; it can, therefore, be concluded that the assumptions for γ and β parameters were correct (Fig. 3.11).

3.1.3 Kelvin model

The linear mass–spring model for the compression phase illustrated in Section 3.1.1, with possibly its extension to the restitution phase illustrated in paragraph Section 3.1.2, is widely used in the reconstruction and analysis

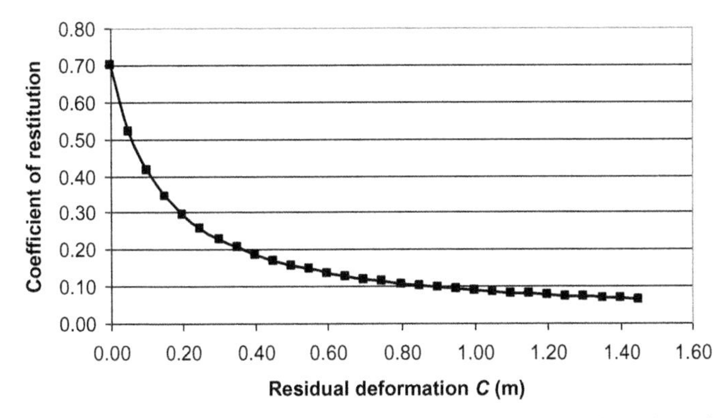

FIGURE 3.11 Coefficient of restitution as a function of the residual deformation, obtained by the sample data.

of road accidents due to its simplicity and the reduced number of parameters needed to apply it.

In the linear mass–spring model, we have a halfsine transient displacement with a centroid time (t_m) of 0.64% of the time of dynamic crush (t_X). In production vehicles the typical range of time is between 46% and 57%, less than that of a spring–mass model (64%). The differences between the test and the spring model are due to the absence of damping in the model.

With the introduction of a damping element in the model the Kelvin models are obtained, in which spring and damper are in parallel, or of Maxwell, in which spring and damper are in series.

The Kelvin model provides the necessary configuration for the vehicle-to-vehicle, vehicle-to-barrier, and component impact modeling, whereas the Maxwell one provides special response characteristics and is suitable for component relaxation and creep modeling and localized vehicle impact modeling. Here, only the Kelvin model is presented, more generally for the analysis of road accidents.

Vehicle-to-barrier Kelvin model

The vehicle is schematized as a mass with spring and damper in parallel, as shown in Fig. 3.12.

The model is defined by the spring stiffness k, the damping coefficient c, and the mass m.

The equation of motion for the Kelvin model has the following form:

$$m\ddot{x} + c\dot{x} + kx = 0 \tag{3.49}$$

assuming

$$\begin{aligned} \zeta &= \frac{c}{2m\omega} \\ \omega &= 2\pi f = \sqrt{\frac{k}{m}} \end{aligned} \tag{3.50}$$

become

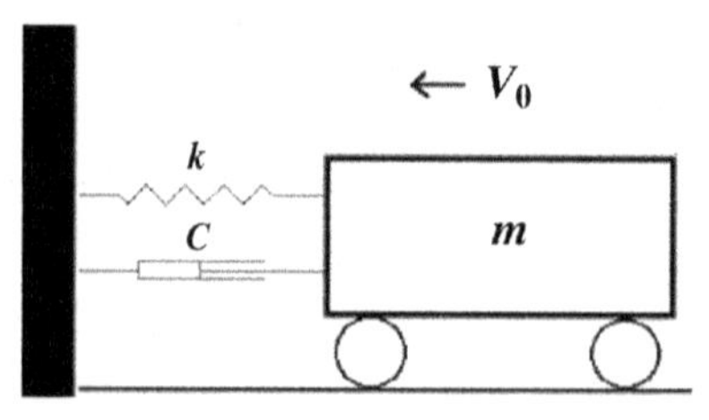

FIGURE 3.12 A vehicle Kelvin model.

$$\ddot{x} + 2\zeta\omega\dot{x} + \omega^2 x = 0 \tag{3.51}$$

whose characteristic equation, applying the Laplace transformation, results

$$s^2 + 2\zeta\omega s + \omega^2 = 0 \tag{3.52}$$

with roots

$$\begin{aligned} s_1 &= a + ib \\ s_2 &= a - ib \end{aligned} \tag{3.53}$$

where

$$\begin{aligned} a &= \zeta\omega \\ b &= \beta\omega \\ \beta &= \sqrt{1-\zeta^2} \end{aligned} \tag{3.54}$$

The general solution of Eq. (3.51) is

$$x = e^{at}[c_1 \sin(bt) + c_2 \cos(bt)] \tag{3.55}$$

with c_1 and c_2 constants to be defined based on the initial displacement and velocity conditions.

Underdamped system

For $0 \leq \zeta < 1$ the system is underdamped, and given V the initial speed and 0 the initial displacement, we have

$$x(t) = \frac{Ve^{-\zeta\omega t}}{\sqrt{1-\zeta^2}\omega} \sin\left(\sqrt{1-\zeta^2}\omega t\right) \tag{3.56}$$

$$\dot{x}(t) = Ve^{-\zeta\omega t}\left[\cos\left(\sqrt{1-\zeta^2}\omega t\right) - \frac{\zeta}{\sqrt{1-\zeta^2}} \sin\left(\sqrt{1-\zeta^2}\omega t\right)\right] \tag{3.57}$$

$$\ddot{x}(t) = V\omega e^{-\zeta\omega t}\left[-2\zeta\cos\left(\sqrt{1-\zeta^2}\omega t\right) + \frac{2\zeta^2-1}{\sqrt{1-\zeta^2}} \sin\left(\sqrt{1-\zeta^2}\omega t\right)\right] \tag{3.58}$$

The displacement pulse shape $x(t)$ of the underdamped system is determined by the product of exponential decayed and sinusoidal function.

The restitution coefficient can be calculated considering that at the end of the impact, $t = t_f$ must be $\ddot{x}(t_f) = 0$ and from Eq. (3.58), we have

$$\tan\left(\sqrt{1-\zeta^2}\omega t_f\right) = \frac{2\zeta\sqrt{1-\zeta^2}}{2\zeta^2-1} \tag{3.59}$$

and from the Pythagorean theorem

$$\begin{aligned} \cos\left(\sqrt{1-\zeta^2}\omega t_f\right) &= 2\zeta^2 - 1 \\ \sin\left(\sqrt{1-\zeta^2}\omega t_f\right) &= 2\zeta\sqrt{1-\zeta^2} \end{aligned} \tag{3.60}$$

The restitution coefficient ε then is

$$\varepsilon = -\frac{\overline{V}}{V} = -\frac{\dot{x}(t_f)}{V} = -e^{-\zeta\omega t_f}\left[\cos\left(\sqrt{1-\zeta^2}\omega t_f\right) - \frac{\zeta}{\sqrt{1-\zeta^2}}\sin\left(\sqrt{1-\zeta^2}\omega t_f\right)\right] \tag{3.61}$$

and substituting Eq. (3.60) in Eq. (3.61), we get

$$\varepsilon = e^{-\zeta\omega t_f} \tag{3.62}$$

For $\zeta = 0$, we obtain $\varepsilon = 1$.

Critically damped system

For ζ that tends to 1, ωt tends to 0, and the system is critically damped and has

$$x(t) = Vte^{-\omega t} \tag{3.63}$$

$$\dot{x}(t) = V(1 - \omega t)e^{-\omega t} \tag{3.64}$$

$$\ddot{x}(t) = V\omega(\omega t - 2)e^{-\omega t} \tag{3.65}$$

The displacement pulse shape $x(t)$ of the critically damped is determined by the product of exponential decayed and time function Vt. Also, in this case, the restitution coefficient can be evaluated. Since $\ddot{x}(t_f) = 0$ from Eq. (3.65), we have $\omega t_f = 2$ and therefore

$$\varepsilon = -\frac{\dot{x}(t_f)}{V} = -(1 - \omega t_f)e^{-\omega t_f} = e^{-2} = 0.135 \tag{3.66}$$

Overdamped system

For $\zeta > 1$ the system is overdamped, and the solutions are formally equal to Eqs. (3.56)–(3.58), in which the trigonometric functions are replaced with the homologous hyperbolic trigonometric functions. The displacement pulse shape $x(t)$ of the critically damped is determined by the product of exponential decayed and hyperbolic sine function.

The restitution coefficient, for $\zeta \to \infty$ tends to 0.

If the elastic loading and unloading stiffness of the spring is the same, the energy dissipation (hysteresis energy) during impact is attributed to damping only. In the real vehicle impacts, plastic unloading occurs, and the computed coefficient of restitution based on the elastic model is higher than the actual one. The underdamped system, with $0 \leq \zeta < 1$, to which corresponds $0 \leq \varepsilon < 0.135$ describes the most real impacts.

Responses of an underdamped systems

The significant responses of an underdamped system are shown in the following.

At the time of dynamic crash t_X the velocity is 0, $\dot{x}(t_x) = 0$, then

$$\frac{1}{\tan\left(\sqrt{1-\zeta^2}\omega t_X\right)} = \frac{\zeta}{\sqrt{1-\zeta^2}} \tag{3.67}$$

and, from Pythagorean theorem

$$\sin\left(\sqrt{1-\zeta^2}\omega t_X\right) = \sqrt{1-\zeta^2} \tag{3.68}$$

The undamped maximum crush, from Eq. (3.56), is

$$X = x(t_X) = \frac{V}{\omega} e^{-\zeta\omega t_x} \tag{3.69}$$

The time at which there is the maximum deformation, placing $\dot{x}(t_X) = 0$ results

$$t_X = \frac{1}{\omega\sqrt{1-\zeta^2}} \arctan\left(\frac{\sqrt{1-\zeta^2}}{\zeta}\right) \tag{3.70}$$

the centroid time results

$$t_C = \frac{X}{V} = \frac{1}{\omega} e^{-\zeta\omega t_X} \tag{3.71}$$

the time of separation velocity (final time) is

$$t_f = 2t_x \tag{3.72}$$

At the time of separation velocity the acceleration is 0, $\ddot{x}(t_f) = 0$, then from Eq. (3.58), we get

$$\tan\left(\sqrt{1-\zeta^2}\omega t_f\right) = \frac{2\zeta\sqrt{1-\zeta^2}}{2\zeta^2 - 1} \tag{3.73}$$

and, from Pythagorean theorem

$$\sin\left(\sqrt{1-\zeta^2}\omega t_f\right) = 2\zeta\sqrt{1-\zeta^2} \tag{3.74}$$

replacing in Eq. (3.56) the residual deformation is obtained at the end of impact C:

$$C = x(t_f) = \frac{V}{\omega} 2\zeta e^{-\zeta\omega t_f} \tag{3.75}$$

The rebound velocity is

$$\overline{V} = \dot{x}(t_f) = -Ve^{-2\zeta\omega t_X} \tag{3.76}$$

Kelvin model parameters identification

The Kelvin model parameters k and c can be estimated from the analysis of a dynamic response or crash pulse of a vehicle from a test. To obtain the parameters, it is necessary to have the time of dynamic crush t_X, initial impact velocity V, dynamic crush X, and mass of the vehicle m.

From Eqs. (3.71) and (3.70), we get relative centroid location:

$$\frac{t_c}{t_X} = \frac{\sqrt{1-\zeta^2}}{\arctan\left(\left(\sqrt{1-\zeta^2}\right)/\zeta\right)} e^{\left[\left(-\zeta/\left(\sqrt{1-\zeta^2}\right)\right)\arctan\left(\left(\sqrt{1-\zeta^2}\right)/\zeta\right)\right]} \tag{3.77}$$

Knowing the relative centroid location by determining t_c and t_x from the crash pulse from the above equation, we get the damping factor ζ.

From Eq. (3.70), we get the structural natural frequency $f = \omega/2\pi$:

$$f = \frac{1}{2\pi t_x\sqrt{1-\zeta^2}} \arctan\left(\frac{\sqrt{1-\zeta^2}}{\zeta}\right) \tag{3.78}$$

Once deriving the damping factor ζ and the natural frequency f, the Kelvin model parameters k and c can be estimated from Eq. (3.50):

$$\begin{aligned} k &= 4\pi^2 f^2 m \\ c &= 4\pi f \zeta m \end{aligned} \tag{3.79}$$

Vehicle-to-vehicle Kelvin model

The vehicle-to-vehicle impact can be schematized as shown in Fig. 3.13, in which each vehicle is schematized with a Kelvin model, that is, a mass with spring elements and damper in parallel.

The two sets of Kelvin elements can be represented as a single set as shown in Fig. 3.14

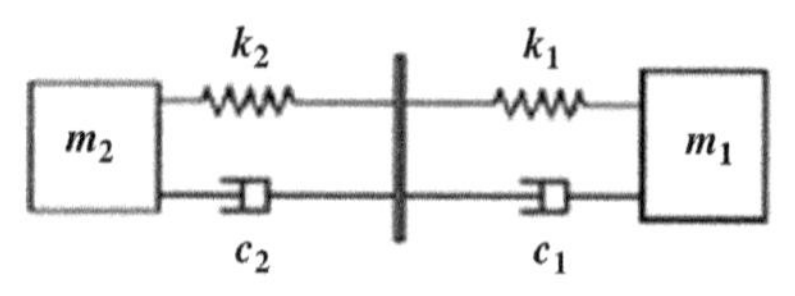

FIGURE 3.13 Vehicle-to-vehicle impact model, represented by two Kelvin elements in series.

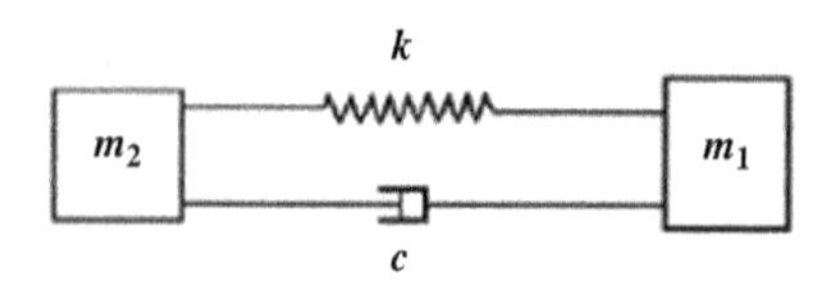

FIGURE 3.14 Vehicle-to-vehicle impact model, represented by an equivalent Kelvin model.

The total deformation x of the springs is equal to the sum of the deformations of the two single springs in series:

$$x = x_1 + x_2 \tag{3.80}$$

that, given the relationship between strength and stiffness, considering only the force passing through the F_k springs, can be written as

$$\frac{F_k}{k} = \frac{F_k}{k_1} + \frac{F_k}{k_2} \tag{3.81}$$

from which the stiffness of the spring equivalent to the two in series results:

$$k_c = \frac{k_1 k_2}{k_1 + k_2} \tag{3.82}$$

Deriving respect to the time Eq. (3.80), we have

$$\dot{x} = \dot{x}_1 + \dot{x}_2 \tag{3.83}$$

that, considering only the force passing through the F_c dampers, can be written as

$$\frac{F_c}{c} = \frac{F_c}{c_1} + \frac{F_c}{c_2} \tag{3.84}$$

from which the damping coefficient equivalent to the two series dampers is

$$c_c = \frac{c_1 c_2}{c_1 + c_2} \tag{3.85}$$

It results $F_k + F_c = F$.

At this point the description of the vehicle-to-vehicle impact can be made with the same equations used for the vehicle-to-barrier impact, using for k and c in Eqs. (3.82) and (3.85), respectively.

3.2 Pulse models

The acceleration during the collision can be approximated through a suitable closed form function, which describes the trend macroscopically.

This is a direct way of characterizing the crash pulse. Accelerometer data relating to different points of the vehicle generally are obtained from experimental crash tests. During the vehicle deformation the structures geometry change; each point of the deformed area has a value of deceleration, and the center of gravity changes its position. However, in a first approximation, one can consider the center of gravity fixed with respect to the undeformed part and, as representative of the center of gravity deceleration, consider the average deceleration between the various accelerometer readings.

A crash pulse model allows the representation of crash pulse time histories, maintaining as many response parameters as possible, such as maximum dynamic crush, velocity change, time of dynamic crush, centroid time, static crush, separation (rebound) velocity, transient acceleration, velocity, displacement in time domain, and energy absorption.

The pulse models are generally used to supplement full-scale dynamic testing of vehicle crashworthiness. However, the pulse models can be conveniently used in accident reconstruction to have more information on the impact phases. As an example, the models can be used to have information on crash acceleration history and crash severity, on injury risk and to verify the correct activation of passive safety devices such as airbags.

In the automotive safety field, several crash pulse approximations and techniques have been developed (Huang, 2002), such as Tipped Equivalent Square Wave, Fourier Equivalent Wave, Trapezoidal Wave Approximation, Bi or Piecewise Slope Approximation (Wei et al., 2017) and basic harmonic pulses. The latter, as haversine, halfsine, and triangular, are the most used pulse functions in the field of accident reconstruction, each of them more suitable for some defined crash configurations. Numerous factors can affect the characteristics of a crash pulse. Some of these include the vehicle shape, vehicle structure, vehicle mass, collision partner, crash mode, and amount of engagement. Different crash modes with the same vehicle can exhibit different collision pulse shapes. Often, the full engagement impacts are better represented by the halfsine model, while the offset impact by the haversine or triangular model (Varat and Husher, 2003).

In the following the generalized models for basic harmonic and triangular pulses are derived for a two-vehicle collision (see Fig. 3.15). The models define the collision pulses through the crash duration T and peak acceleration P. These parameters require the knowledge of initial speed V, change in velocity ΔV, and crush data C, usually obtained from the reconstruction of the impact phase of the accident. Once obtained the pulse, the outputs of the

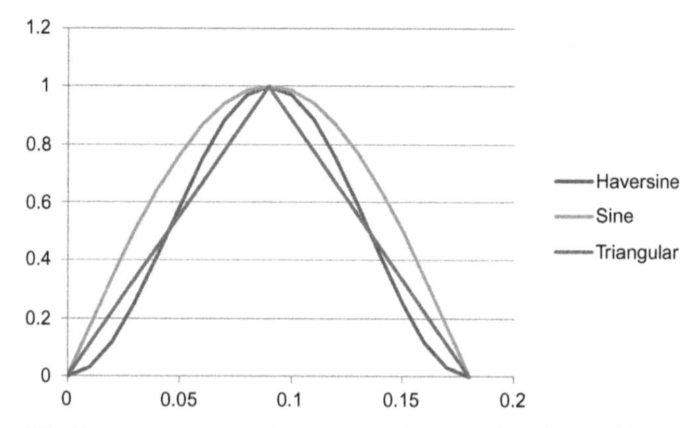

FIGURE 3.15 Haversine, sine, and triangular approximations of acceleration/time response.

models include the acceleration, velocity, and displacement–time histories, the peak acceleration, average acceleration, and impact duration.

3.2.1 Halfsine pulse shape

In this model the acceleration can be approximated by the sin function:

$$a(t) = P\sin\left(\frac{\pi}{T}t\right) \tag{3.86}$$

where T is the pulse duration and P is the acceleration peak value.

The velocity can be obtained by integrating the acceleration function, with the initial condition $V(0) = V$:

$$V(t) = \frac{TP}{\pi}\left[1 - \cos\left(\frac{\pi}{T}t\right)\right] + V \tag{3.87}$$

Integrating Eq. (3.87), with the initial condition $x(0) = 0$, yields the displacement:

$$x(t) = \frac{TP}{\pi}\left[t - \frac{T}{\pi}\sin\left(\frac{\pi}{T}t\right)\right] + tV \tag{3.88}$$

To determine the P and T parameters, starting from speed V, change in velocity ΔV, and crush data C, different methodologies can be used:

1. The collision duration T between two vehicles, in case of a full frontal impact, is determined considering the mutual crush $= x_1(T) - x_2(T)$. Substituting these conditions and Eq. (3.98) into the displacement–time equation yields the following equation for the duration, T:

$$T = \frac{|x_1(T) - x_2(T)|}{\left|\left(V_1 + (\Delta V_1/2)\right) - \left(V_2 + (\Delta V_2/2)\right)\right|} \tag{3.89}$$

It is noted that Eq. (3.89) represents the relationship between the relative space traveled during the impact and the average relative speed. However, the space $x(T)$ is relative to the compression phase plus the elastic restitution, so that for impacts with nonnegligible elastic restitution, $x_1(T) - x_2(T)$ does not coincide with the difference between the residual deformations $C_1 - C_2$. In these cases, it is more correct to consider the time of the compression and return phase separately.

The speed variation during the compression phase is linked to the one during the rebound phase from the restitution coefficient:

$$\Delta V_r = \varepsilon \Delta V_c \tag{3.90}$$

and, as we also have

$$\Delta V_r + \Delta V_c = \Delta V \tag{3.91}$$

we get

$$\Delta V_c = (1 + \varepsilon)\Delta V \tag{3.92}$$

Then the duration of the compression phase results

$$T_c = \frac{2(1 + \varepsilon)|x_1(T) - x_2(T)|}{|\Delta V_1 - \Delta V_2|} \tag{3.93}$$

Considering that the deformation in the rebound phase is equal to $x(T) - C$, the duration of the rebound phase is

$$T_c = \frac{2(1 + \varepsilon)|(x_1(T) - C_1) - (x_2(T) - C_2)|}{\varepsilon|\Delta V_1 - \Delta V_2|} \tag{3.94}$$

from which the total impact time is $T = T_c + T_r$.

In the event of a collision with a rigid barrier, Eq. (3.93) becomes

$$T_c = \frac{2X}{V} \tag{3.95}$$

and Eq. (3.94)

$$T_r = \frac{2(x(T) - C)}{\Delta V - V} \tag{3.96}$$

Generally, from the crash test analysis, the duration of the restitution phase is about 40% of that of the compression phase, so the overall duration can also be approximated as

$$T = \frac{2.8(1 + \varepsilon)|x_1(T) - x_2(T)|}{|\Delta V_1 - \Delta V_2|} \tag{3.97}$$

Once determined the impact duration T, the peak acceleration P can be obtained from Eq. (3.87) considering that $V(T) = V + \Delta V$:

$$P = \frac{\Delta V \pi}{2T} \tag{3.98}$$

2. In the event of a nonfull frontal impact, that is, when the deformation does not take on a "rectangular" shape (see Section 4.6), the duration of the impact can no longer be evaluated using the methods described in point 1, because the entity of the mutual crash varies from point to point and its direction could be not aligned with that of speed. In this case, it is possible to apply the triangle method (see Section 4.6) to calculate the maximum force exchanged during the impact.

   Referring to Section 4.7, the force in the direction parallel to the vehicle axis acting on the section dl is

$$dF = Fdl = Bxdl \tag{3.99}$$

and, as (Fig. 3.16)

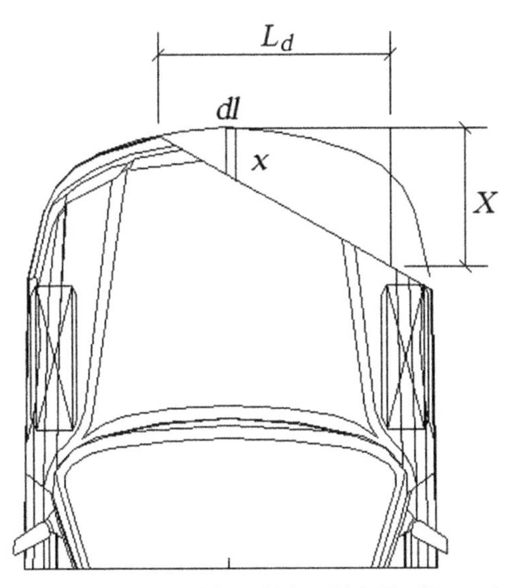

FIGURE 3.16 Linearized deformation of the vehicle, which identifies a triangular area.

$$x = \frac{X}{L_d} l \tag{3.100}$$

the resulting force over the entire deformed area is.

$$F_{\max} = \int_0^{L_d} \frac{BX}{L_d} l dl = \frac{BXL_d}{2} \tag{3.101}$$

Dividing Eq. (3.101) by mass m gives the maximum acceleration parallel to the vehicle axis $P_{\backslash\backslash}$:

$$P = \frac{BXL_d}{2m} \tag{3.102}$$

The peak acceleration P can be obtained from Eq. (3.102), considering the principal direction of force (PDOF):

$$P = \frac{BXL_d}{2\cos(\text{PDOF})} \tag{3.103}$$

From Eq. (3.98), we can, therefore, derive the duration of the impact:

$$T = \frac{\Delta V \pi}{2P} \tag{3.104}$$

3. The parameters of the acceleration curve can also be evaluated considering the instant t_X corresponding to the maximum compression between the vehicles, that is, the maximum dynamic deformations. When the maximum compression occurs, the speed of the two vehicles is the same.

The time t_X that has the maximum dynamic deformation X can, therefore, be obtained from Eq. (3.87) by imposing $V_1(t_X) = V_2(t_X)$:

$$\frac{TP_1}{\pi}\left[1-\cos\left(\frac{\pi}{T}t_X\right)\right]+V_1=\frac{TP_2}{\pi}\left[1-\cos\left(\frac{\pi}{T}t_X\right)\right]+V_2 \tag{3.105}$$

Introducing the variable y

$$y=\frac{t_x}{T} \tag{3.106}$$

and considering Eq. (3.98), we get

$$y=\frac{1}{\pi}\arccos\left[1+\frac{2(V_1-V_2)}{\Delta V_1-\Delta V_2}\right] \tag{3.107}$$

Since at time t_X must be $x_1(t_X)-x_2(t_X)=X_1-X_2$, from Eqs. (3.88), (3.105), and (3.98), we get the impact duration:

$$T=\frac{X_1-X_2}{(\Delta V_1-\Delta V_2)/2\left[y-(1/\pi)\sin(\pi y)+y(V_1-V_2)\right]} \tag{3.108}$$

Once determined the impact duration T, the peak acceleration P can be obtained from Eq. (3.87) considering that at $V(T)=V+\Delta V$:

$$P=\frac{\Delta V\pi}{2T} \tag{3.109}$$

3.2.2 Haversine pulse shape

In this model the acceleration can be approximated by the $\sin^2$ function:

$$a(t)=P\sin^2\left(\frac{\pi}{T}t\right) \tag{3.110}$$

where T is the pulse duration and P is the acceleration peak value.

The velocity can be obtained by integrating the acceleration function, with the initial condition that for $t=0$, $V(0)=V$:

$$V(t)=\frac{P}{4\pi}\left[2\pi t-T\sin\left(\frac{2\pi}{T}t\right)\right]+V \tag{3.111}$$

Integrating Eq. (3.113), with the initial condition $t=0$, $x(0)=0$, yields the displacement:

$$x(t)=\frac{T^2P}{8\pi^2}\left[\cos\left(\frac{2\pi}{T}t\right)-1\right]+\frac{t}{4}(Pt+4V) \tag{3.112}$$

To determine the parameters P and T, starting from closing speed V, change in velocity ΔV, and crush data X, different methodologies can be used, similarly to what was done for the halfsine pulse shape:

1. The collision duration T between two vehicles, in case of a full frontal impact, can be determined considering the mutual crush $= x_1(T) - x_2(T)$, as done for the halfsine pulse, Eqs. (3.89), (3.93), (3.94), and (3.97). Then, the peak acceleration P can be obtained from Eq. (3.111) considering that at $t = T$, $V = V + \Delta V$:

$$P = \frac{2\Delta V}{T} \tag{3.113}$$

2. In the event of a nonfrontal collision the pulse peak can be determined by the same methodology as for the halfsine pulse. However, in this case, there is an approximation, because the force no longer has a linear trend with the deformation x, as for the halfsine pulse model, and Eq. (3.99) is no longer strictly valid. With this approximation, P can be obtained from Eq. (3.103), then the impact duration can be obtained from Eq. (3.113)

$$T = \frac{2\Delta V}{P} \tag{3.114}$$

3. Similarly to what was done for the halfsine pulse, the parameters of the acceleration curve can be evaluated considering the moment t_X at which the maximum dynamic deformations of the vehicles occur. When the maximum compression occurs, the speed of the two vehicles is the same. The time t_X with the maximum dynamic deformation X can, therefore, be obtained from Eq. (3.111) by imposing $V_1(t_X) = V_2(t_X)$:

$$\frac{P_1}{4\pi}\left[2\pi t_x - T\sin\left(\frac{2\pi}{T}t_X\right)\right] + V_1 = \frac{P_2}{4\pi}\left[2\pi t_x - T\sin\left(\frac{2\pi}{T}t_X\right)\right] + V_2 \tag{3.115}$$

Introducing the y variable

$$y = \frac{t_x}{T} \tag{3.116}$$

and considering Eq. (3.114), we get

$$y - \frac{1}{2\pi}\sin(2\pi y) = \frac{V_2 - V_1}{\Delta V_1 - \Delta V_2} \tag{3.117}$$

that can be solved numerically to determine y.

Since at time t_X must be $x_1(t_X) - x_2(t_X) = X_1 - X_2$, from Eqs. (3.105), (3.112), and (3.113), we obtain the impact duration:

$$T = \frac{X_1 - X_2}{(\Delta V_1 - \Delta V_2)/4\pi^2[2\pi^2 y^2 - 1 + \cos(2\pi y)] + y(V_1 - V_2)} \tag{3.118}$$

3.2.3 Triangular pulse shape

In this model the acceleration can be approximated by the triangular function defined by

$$a(t) = \frac{P}{kT}t \qquad \text{for } 0 \le t \le kT$$
$$a(t) = P\frac{(T-t)}{T(1-k)} \qquad \text{for } kT < t \le T \tag{3.119}$$

where T is the pulse duration and P is the acceleration peak value, occurring at time kT, where $0 < k < T$.

A possible choice of k is to have $kT = t_x$.

To obtain the velocity, Eq. (3.119) can be integrated into two parts. In the first part the integration is for $t \le kT$ with the initial condition that for $V(0) = V$:

$$V(t) = \frac{P}{2kT}t^2 + V \quad \text{for } t \le kT \tag{3.120}$$

Integrating Eq. (3.120), with the initial condition $x(0) = 0$, yields the displacement:

$$x(t) = \frac{P}{6kT}t^3 + V\,t \qquad \text{for } t \le kT \tag{3.121}$$

In the second part, for $t \ge kT$, integrating the second of Eq. (3.119) with the initial condition $V(kT) = \left(PkT/2\right) + V$, the velocity is obtained:

$$V(t) = \frac{-P}{2T(1-k)}t^2 + \frac{PT}{T(1-k)}t + V - \frac{PkT}{2(1-k)} \qquad \text{for } t \ge kT \tag{3.122}$$

Integrating Eq. (3.122), considering the condition that the displacement S at time equal to half the duration must be equal to the value obtained from Eq. (3.121), yields displacement:

$$x(t) = \frac{P}{2(1-k)}\left[-\frac{t^3}{3T} + t^2 - kTt + \frac{k^2T^2}{3}\right] + V\,t \qquad \text{for } t \ge kT \tag{3.123}$$

To determine the parameters P and T, starting from initial speed V, change in velocity ΔV, and crush data X, different methodologies can be used, similarly to what was done for the halfsine and haversine pulse shape:

1. The collision duration T between two vehicles, in case of a full frontal impact, can be determined considering the mutual crush as done for the halfsine pulse, Eqs. (3.89), (3.93), (3.94), and (3.97). Then the peak acceleration P can be obtained from Eq. (3.122) considering that $V(T) = V + \Delta V$:

$$P = \frac{2\Delta V}{T} \tag{3.124}$$

2. In the event of a nonfrontal collision the pulse peak can be determined by the same methodology as the halfsine and haversine pulse, Eq. (3.103), then the impact duration can be obtained from Eq. (3.124):

$$T = \frac{2\Delta V}{P} \tag{3.125}$$

3. Similarly to what was done for the halfsine and haversine pulse, the parameters of the acceleration curve can be evaluated considering the moment t_X at which the maximum dynamic deformations of the vehicles occur. When the maximum compression occurs, the speed of the two vehicles is the same. The time t_X with the maximum dynamic deformation X can, therefore, be obtained from Eq. (3.120) by imposing $V_1(t_X) = V_2(t_X)$:

$$\frac{P_1 t_x}{2} + V_1 = \frac{P_2 t_x}{2} + V_2 \tag{3.126}$$

Introducing the y variable

$$y = \frac{t_x}{T} = k \tag{3.127}$$

and considering Eq. (3.124), we get

$$y = \frac{V_2 - V_1}{\Delta V_2 - \Delta V_1} \tag{3.128}$$

Since the time t_X must be $x_1(t_X) - x_2(t_X) = X_1 - X_2$, from Eqs. (3.105), (3.121), and (3.124), we get the impact duration:

$$T = \frac{X_1 - X_2}{y^2\left((\Delta V_2 - \Delta V_1)/3\right) + (V_1 - V_2)y} \tag{3.129}$$

3.2.4 Macmillan model

In the approach proposed by Macmillan (1983) to describe the impact behavior of the vehicle, the trend of deceleration over time is schematized through a function as

$$a(t) = -\frac{cV}{T}\tau(1-\tau)^{\beta} \tag{3.130}$$

where $\tau = t/T$ is the dimensionless parameter that corresponds to the time and β a constant between 1 and $(2/\varepsilon - 1)$; the constant c is defined as $c = (1+\varepsilon)(\beta+1)(\beta+2)$. The trend of $a(t)$ is shown in Fig. 3.17.

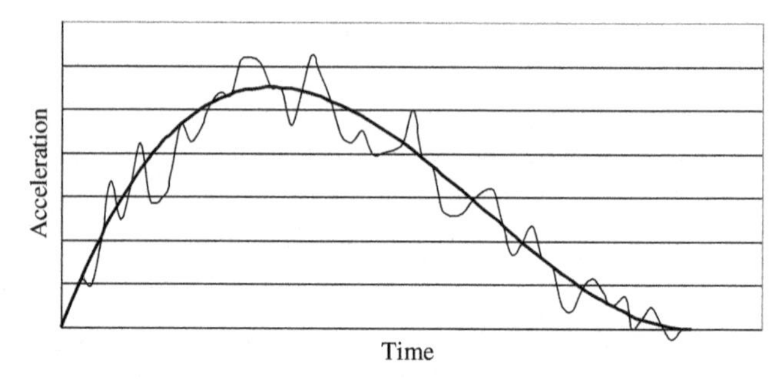

FIGURE 3.17 Trend of experimental acceleration and interpolating functions $a(t)$.

Eq. (3.130) is a time function multiplied by a power curve and is termed an idealized power curve. As the power curve can be represented by binomial series:

$$(1-x)^n = 1 - nx + \frac{n(n-1)x^2}{2!} - \frac{n(n-1)(n-2)x^3}{3!} + \cdots \tag{3.131}$$

and the exponentially decaying curve can be replaced by the power curve function:

$$e^{-x} = 1 - x + \frac{x^2}{2!} - \frac{x^3}{3!} + \frac{x^4}{4!} - \cdots \tag{3.132}$$

the curve used by Macmillan to approximate a vehicle dynamic response has analogies with the Kelvin transient response. In fact, in the Kelvin model, seen in Section 3.1.3, one of the two functions that control the pulse shape is an exponentially decaying function.

The Macmillan curve (3.130) is consistent with the main features that are found in the experimental curves, namely,

1. The impact starts at $t=0$; in this, instant crushing and acceleration are 0, and the speed is equal to V; in mathematical terms, it is required that $a(t)$ passes through the origin.
2. The force increases over time until it reaches a maximum value; the function $|a(t)|$ must, therefore, have an absolute maximum point.
3. The final acceleration is again null because at the end of the impact the vehicle and barrier not exchange more forces; also you must have $da/dt|_{t=T} = 0$, which ensures a gradual end of the impact.

By integrating $a(t)$ and imposing the boundary condition $V(0) = V$, we get

$$V(t) = V\left\{(1+\varepsilon)[1+(\beta+1)\tau](1-\tau)^{\beta+1} - \varepsilon\right\} \tag{3.133}$$

By integrating it again and imposing that $x(0) = 0$, it has

$$x(t) = VT\left\{\frac{1+\varepsilon}{\beta+3}\left[2-(2+(\beta+1)\tau)(1-\tau)^{\beta+2}\right]-\varepsilon\tau\right\} \tag{3.134}$$

Of particular importance is the value of residual deformation or crushing:

$$x(T) = C = V\,T\left[\frac{2(1+\varepsilon)}{\beta+3}-\varepsilon\right] \tag{3.135}$$

Using this acceleration model, we can find three parameters that completely describe the response of the vehicle during an impact.

One can assume that for ε values close to 0 β can be written as

$$\beta = \beta_0 - (\beta_0 - 1)\varepsilon \tag{3.136}$$

Note that, even if Eq. (3.136) is written for $\varepsilon \cong 0$, it complies with the condition for which $\beta = 1$ for $\varepsilon = 1$.

The β_0 parameter is the structural index. Since $\beta > 1$, also $\beta_0 > 1$; a typical value of β_0 is around 2. The second parameter that is introduced is the crushing module, analogous to the stiffness constant of a spring is defined as the initial slope of the curve *F(x)*:

$$K = \frac{dF}{dx}\bigg|_{t=0} = m\frac{da}{dx}\bigg|_{t=0} = \frac{m(da/dt)_{t=0}}{dx/dt|_{t=0}} \tag{3.137}$$

It is easy to see that

$$\frac{dx}{dt}\bigg|_{t=0} = V\frac{da}{dt}\bigg|_{t=0} = \frac{cV}{T^2} \tag{3.138}$$

from which

$$K = \frac{mc}{T^2} \tag{3.139}$$

The typical values of this constant K are around $1.2\cdot 10^6\ N/m$.

The average force exchanged during the impact is given by

$$F_{av} = \frac{m(V-\overline{V})}{T} = \frac{mV(1+\varepsilon)}{T} \tag{3.140}$$

Varying the coefficient of restitution, it is seen that the curve (F_{av}, ε) has the trend of a negative exponential (see Fig. 3.27), in other words, we can write (Fig. 3.18)

$$\varepsilon = \exp\left(-\frac{F_{av}}{K_r}\right) \tag{3.141}$$

The K_r quantity is referred to as "impact resistance factor" and is the third and last parameter; like the other two constant defined earlier, it is a

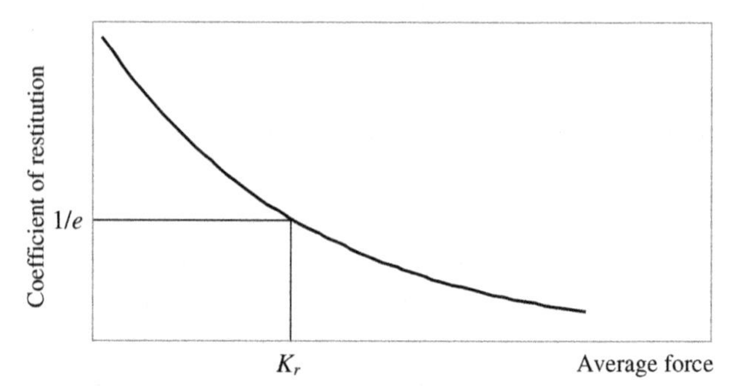

FIGURE 3.18 Trend of $\varepsilon(F_{av})$ function.

characteristic of the vehicle and can be calculated by replacing Eq. (3.140) to Eq. (3.141) and obtaining

$$K_r = \frac{mV}{T} \frac{1+\varepsilon}{\ln(1/\varepsilon)} \tag{3.142}$$

A typical value is $K_r = 65$ kN.

There are two particular moments: t_m, wherein the force exchanged is maximum, and t_X, in which there is maximum deformation. The first value is obtainable in an explicit form by placing equal to zero the first derivative of $a(t)$, from which it follows that

$$t_m = T\frac{1}{\beta+1} \tag{3.143}$$

The value t_X cannot instead be expressed analytically and has to be found numerically putting $V(t) = 0$.

It is possible to extend the acceleration model of Macmillan used to describe the vehicle impact against the barrier, to the case of frontal impact between two vehicles. The assumptions made are as follows:

1. The two colliding vehicles are schematized as impacting against a massless barrier (mobile), placed on the contact surface (see Fig. 3.19).
2. The characteristic $F(x)$ of each vehicle is identical to that which occurs in the collision against the barrier that produces the same residual deformation.

This second hypothesis does not imply that the function $F(t)$ is the same for the two impacts, because in the collision against barrier, we have $F = m\ddot{x}$, while in the case of two vehicles we have $F = m(\ddot{x} - \ddot{x}_p)$, where $\ddot{x}_p$ is the acceleration of the ideal barrier that one imagines at the contact surface, which is generally not fixed. However, for the principle of action and reaction, $F_A(t) = F_B(t)$.

To mathematically describe this event, consider the closing speed between vehicles: $V_R = V_A + V_B$.

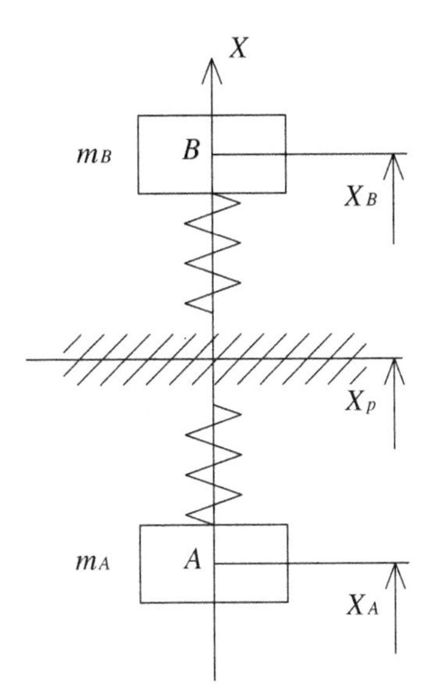

FIGURE 3.19 Model for the collision between two vehicles.

It is then assumed that, similarly to what was done for the collision with the barrier, we can write

$$\dot{V}_R = \frac{(1+\varepsilon_c)(\beta_c+1)(\beta_c+2)V_{R1}}{T}(\tau)(1-\tau)^{\beta_c} \tag{3.144}$$

To highlight the fact that all the constants that appear are different (or at least, not necessarily equal) to those of impact against the barrier, they are indicated with the subscript c (for collision). By integrating this expression over time, it can be found expressions similar to those for the impact against the barrier, provided we replace

$$\begin{aligned}
&C_c = C_A + C_B && \text{instead of } C; \\
&V_R && \text{instead of } V; \\
&m_c = \frac{m_A m_B}{m_A + m_B} && \text{instead of } m \\
&K_c = \frac{K_A K_B}{K_A + K_B} && \text{instead of } K
\end{aligned} \tag{3.145}$$

Suppose, we know the structural characteristics $F(x)$ of the two vehicles and the residual crush of one of the two, for example, the vehicle A.

Since it is assumed that the characteristic $F(x)$ for the vehicle A is the same that would occur in the collision against the barrier that causes C_A, one can determine the maximum force against the barrier and then use it in the collision between two vehicles. In particular, by combining the definitions of K, and the K_r and Eq. (3.135), we get

$$C_A \frac{K_A}{K_{rA}} = \ln\left(\frac{1}{\varepsilon_A}\right) \frac{(\beta_A + 1)(\beta_A + 2)[2(1 + \varepsilon_A) - (\beta_A + 3)\varepsilon_A]}{\beta_A + 3} \tag{3.146}$$

where ε_A and β_A are the coefficients relating to the impact against the barrier that generates the same deformation C_A.

Recalling that, knowing β_{0A}, β_A is a function only of ε_A that is the only unknown of this equation, which can be solved numerically. Obtaining ε_A, it is possible to calculate the maximum force exchanged in the collision that proves to be

$$F_{\max} = K_{rA} \ln\left(\frac{1}{\varepsilon_A}\right)(\beta_A + 2)\left(\frac{\beta_A}{\beta_A + 1}\right)^{\beta_A} \tag{3.147}$$

The same equation can be written using the variables relating to vehicle B; since then β_B is function of ε_B, this quantity can be obtained by solving the numerically equation. From the definition of K_r and from Eq. (5.39), one can then derive V_{B1} and s_{sB}.

Knowing all these parameters, one can set the system of three independent equations:

$$\begin{cases} F_{\max} = \dfrac{C_c K_c \beta_c^{\beta_c}(\beta_c + 3)}{(\beta_c + 1)^{\beta_c + 1}[2 - \varepsilon_c(\beta_c + 1)]} \\ \Delta E = \dfrac{1}{2} m_A V_A^2 (1 - \varepsilon_A^2) + \dfrac{1}{2} m_B V_B^2 (1 - \varepsilon_B^2) = \dfrac{1}{2} m_c V_R^2 (1 - \varepsilon_c^2) \\ F_{\max} = K_c V_R \dfrac{\beta_c^{\beta_c}}{(\beta + 1)^{\beta_c + 1}} \sqrt{\dfrac{m_c}{K_c}(1 + \varepsilon_c)(\beta_c + 1)(\beta_c + 2)} \end{cases} \tag{3.148}$$

The system can be solved numerically to derive the three unknowns ε_c, β_c, and V_R. If it is known V_R instead *of* C_A, it is necessary to apply an iterative process; it is assumed an attempt value of C_A the calculations set out above should be repeated until we find the V_R value corresponding to the first attempt. Comparing this value with the real one, it is possible to establish a second attempt of C_A (taking into account that V_R and C_A are in first approximation directly proportional), and so on, until an acceptable convergence is reached.

Finally, we consider how we should correct this model in the case of an oblique impact. Macmillan suggests the hypothesis that the three vehicle constants (β_0, K, and K_r) have an elliptical polar pattern, that is, we can write

$$K = \begin{cases} \dfrac{K_f K_l}{\sqrt{K_f^2 \cos^2\vartheta + K_l^2 sin^2\vartheta}} & 0 \text{ degrees} \le \vartheta \le 90 \text{ degrees} \\ \dfrac{K_p K_l}{\sqrt{K_p^2 \cos^2\vartheta + K_l^2 sin^2\vartheta}} & 90 \text{ degrees} \le \vartheta \le 180 \text{ degrees} \end{cases} \tag{3.149}$$

and similarly for β_0 and K_r (the subscript f stands for "front," l for "lateral," and p for "rear"). It must also be considered that the resultant of the forces exchanged not necessarily passes through the centers of mass of the two vehicles, which gives rise to a moment along the yaw axis direction. In this case, in place of the actual mass of the vehicle, it must be used the mass reduced by the γ factor, already seen in Eq. (2.45).

3.2.5 Example

Consider the crash test frontal oblique NHTSA no. CB0300 of the Jeep Grand Cherokee 2011 vehicle, against a rigid 30 degrees inclined barrier, of which the following data are known:

• Impact speed	39.9 km/h
• Impact duration	188 ms
• ΔV	47.5 km/h
• Vehicle mass	2360 kg

We want to approximate the acceleration, velocity, and displacement curves over time, applying the approximations that are seen earlier.

The deformation of the vehicle, using the triangle method, can be estimated as having a triangular geometry and a maximum depth of 0.50 m, as illustrated in Fig. 3.20.

The dynamic deformation X results

$$X = C + \frac{b_0}{b_1} = 0.58 \text{ m}$$

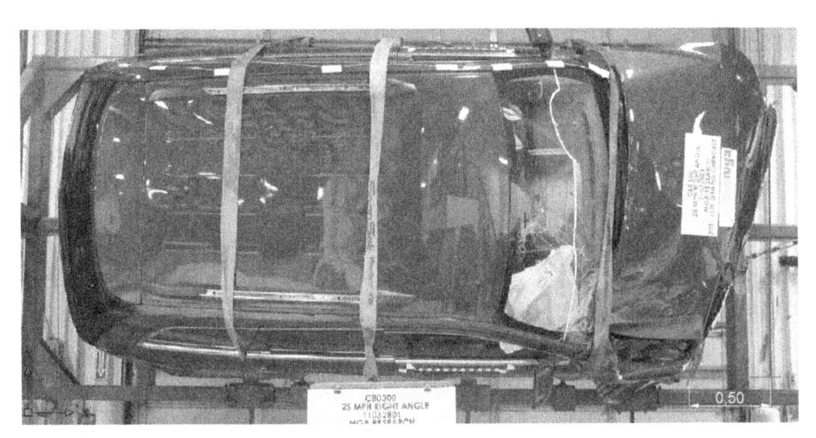

FIGURE 3.20 Deformation of the crashed vehicle.

while the PDOF results

$$\text{PDOF} = \arctan\left(\frac{X}{L_d}\right) = 17$$

$$\Delta V_{//} = \Delta V \cos(\text{PDOF}):\ 45.5\ \text{km}/h$$

The b_1 parameter can be determined using a frontal 100% crash test as a reference model. From the NHTSA crash test no. MB0311, we have the following data:

• Impact speed V	56.3 km/h
• Final speed $\overline{V}$	−5.5 km/h
• ΔV	61.8 km/h
• EES	15.6 m/s
• C	0.52 m
• Frontal width L_{100}	1.944 m
• Damage width L_d	1.944 m

from which we get

$$\varepsilon = -\frac{\overline{V}}{V} = 0.0977$$

$$b_0 = \overline{V} = 1.53\ m/s$$

$$b_1 = \frac{\text{EES} - b_0}{C} = 27\ \text{s}^{-1}$$

To compare the results of the various curve approximation models, from the crash test on the vehicle under analysis, the acceleration curve in the longitudinal direction of the vehicle is available.

Applying the sine pulse shape model, depending on the different methods to obtain the parameters, we have

1. $T = \dfrac{2.8(1+\varepsilon)|X|}{|\Delta V_{//}|} = 168\ \text{ms}$

$$P_{//} = \frac{\Delta V_{//}\pi}{2T} = 12.06\ \text{g}$$

2. $P_{//} = \dfrac{b_1^2 X L_d}{2L_{100}} = 20.7\ \text{g}$

$$T = \frac{\Delta V_{//}\pi}{2P_{//}} = 98\ \text{ms}$$

3. $$y = \frac{1}{\pi}\arccos\left[1 + \frac{2V_0}{\Delta V_{//}}\right] = 0.769$$

$$T = \frac{X}{\Delta V_{//}/2\left[y - (1/\pi)\sin(\pi y) + yV_0\right]} = 112 \text{ ms}$$

$$t_X = \frac{T}{\pi}\arccos\left[1 + \frac{\pi V_0}{TP}\right] = 76 \text{ ms}$$

$$P_{//} = \frac{\Delta V_{//}\pi}{2T} = 18.2 \text{ g}$$

Fig. 3.21 shows the longitudinal acceleration curves of the halfsine pulse shape model for the different parameters calculation methods, compared with the experimentally obtained.

Method 1 estimates a realistic impact duration but a low maximum acceleration. The other two methods, on the other hand, estimate a good maximum acceleration but a low-impact duration.

Applying the haversine pulse shape model, depending on the different methods to obtain the parameters, we have

1. $$T = \frac{2.8(1+\varepsilon)|X|}{|\Delta V_{//}|} = 168 \text{ ms}$$

$$P_{//} = \frac{2\Delta V_{//}}{T} = 15.38 \text{ g}$$

2. $$P_{//} = \frac{b_1^2 X L_d}{2L_{100}} = 20.7 \text{ g}$$

$$T = \frac{2\Delta V_{//}}{P_{//}} = 125 \text{ ms}$$

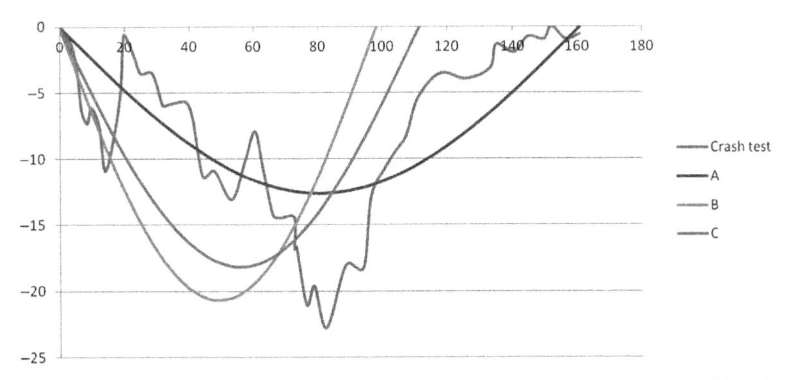

FIGURE 3.21 Approximation of the longitudinal acceleration curve with the halfsine shape model, with the different methods for determining the parameters.

3. Solving numerically the $y-(1/2\pi)\sin(2\pi y)=V_0/\Delta V$, we get $y=0.718$

$$T=\frac{X}{(\Delta V/4\pi^2)[2\pi^2y^2-1+\cos(2\pi y)]+yV_0}=110\ \text{ms}$$

$$t_X=yT=79\ \text{ms}$$

$$P_{//}=\frac{2\Delta V_{//}}{T}=23.65\ \text{g}$$

Fig. 3.22 shows the longitudinal acceleration curves of the haversine pulse shape model, for the different methods of calculating the parameters, compared with the crash test curve.

The haversine shape pulse model seems to better estimate, for all the methods used, the acceleration curve, even if we observe the same type of approximation for methods 1–3.

Applying the triangle pulse shape model, with $kT=t_x$, depending on the different modes to obtain the parameters, we have

1. $T=\dfrac{2.8(1+\varepsilon)|X|}{|\Delta V_{//}|}=161\ \text{ms}$

$$P_{//}=\frac{2\Delta V_{//}}{T}=16.1\ \text{g}$$

$$t_X=\frac{T}{\pi}\arccos\left[1+\frac{\pi V_0}{TP}\right]=75\ \text{ms}$$

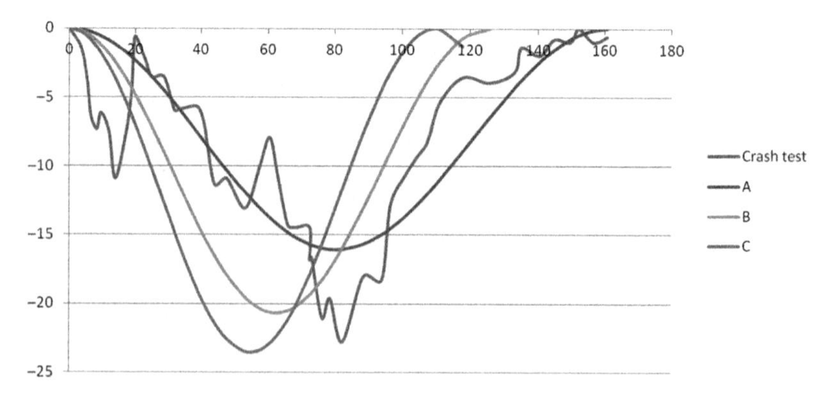

FIGURE 3.22 Approximation of the longitudinal acceleration curve with the haversine shape model, with the different methods for determining the parameters.

2. $P_{//} = \dfrac{b_1^2 X L_d}{2L_{100}} = 20.7\text{ g}$

$$T = \frac{2\Delta V_{//}}{P_{//}} = 125\text{ ms}$$

3. Solving numerically the $y - (1/2\pi)\sin(2\pi y) = (V_0/\Delta V)$, we get $y = 0.718$

$$t_X = yT = 90\text{ ms}$$
$$y = \frac{V_0}{\Delta V} = 0.874$$

$$T = \frac{X}{y^2(\Delta V/3) + V_1 y} = 86\text{ ms}$$
$$t_X = yT = 75\text{ ms}$$
$$P_{//} = \frac{2\Delta V_{//}}{T} = 30.0\text{ g}$$

The longitudinal acceleration curves of the triangle pulse shape model are shown in Fig. 3.23, for the different parameters calculation methods, compared with the crash test curve.

The triangle shape pulse model provides different estimates varying the method used. The best approximation is obtained with method 2.

Applying the Macmillan model, assuming the value of the maximum acceleration

$$P_{//} = \frac{b_1^2 X L_d}{2L_{100}} = 20.7\text{ g}$$

and solving the two equations numerically

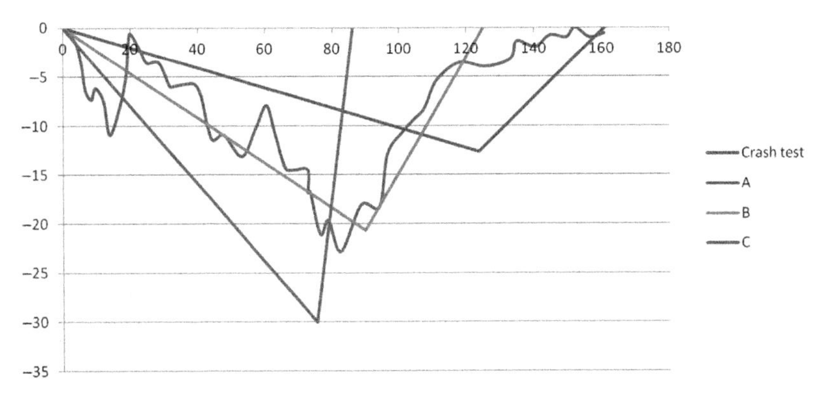

FIGURE 3.23 Approximation of the longitudinal acceleration curve with the triangle shape model, with the different methods for determining the parameters.

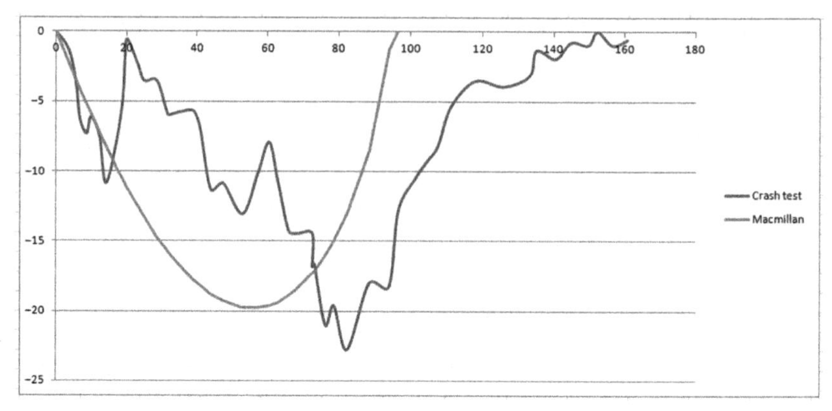

FIGURE 3.24 Approximation of the longitudinal acceleration curve with the Macmillan model.

$$P_{//} = \frac{V_0}{T}(1+\varepsilon)(\beta+2)\left(\frac{\beta}{\beta+1}\right)^{\beta}$$

$$C = V_0 T\left[\frac{2(1+\varepsilon)}{\beta+3} - \varepsilon\right]$$

we get $T = 94$ ms.

The acceleration curve determined with the Macmillan model is reported in Fig. 3.24.

The various models applied show limitations in approximating the acceleration curve of the crash test, partly due to the peculiarity of the experimental curve, which shows a first peak around 15 ms, and then returns to 0 and then up again. This trend is likely due to a structure that after having offered an initial resistance has succumbed. This particular shape of the curve makes it difficult to obtain a ΔV (area subtended by the acceleration curve) and at the same time duration and a maximum acceleration congruent with the real values.

The example, therefore, shows that the various models used are to be understood as broad since often the real behavior presents trends that are not easily traceable to simple geometries.

3.3 Direct integration of the curves *F*(*x*)

A method for the evaluation of the kinematic parameters during the impact consists of carrying out a direct integration of the curves $F(x)$ (Varat and Husher, 2000).

Consider, for simplicity, a one-dimensional collision, as shown in Fig. 3.25.

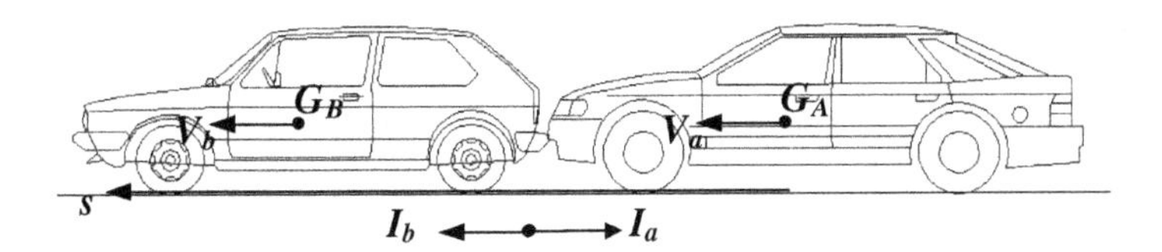

FIGURE 3.25 Diagram of two vehicles in a one-dimensional impact.

The motion of the two vehicles is analyzed dividing the impact duration into short time steps Δt; assuming an attempt value of the initial impact speeds of the vehicles; for each time step, it is possible to calculate the final velocity, writing the equations that express the conservation of momentum, energy, and impulse theorem. The equations are written in a different way for the compression phase and the restitution one, given that in the latter part of the energy is recovered and not absorbed.

Referring to the conditions at the initial time $t=0$, time of maximum deformation t_X, and those at the final time T, we can write

$$\text{Compression}\begin{cases} |I_2| = -|I_1| \\ \int_0^{tx} F(t)dt = |I_1| \\ E_c = \overline{E}_c + E_a \end{cases} \tag{3.150}$$

$$\text{Restitution}\begin{cases} |I_2| = -|I_1| \\ \int_{tx}^{tT} F(t)dt = |I_1| \\ E_c = \overline{E}_c - E_r \end{cases} \tag{3.151}$$

where I is the impulse, E_c is the kinetic energy of the two vehicles, E_a is the energy absorbed, and E_r is the elastically recovered energy.

Make explicit the equations for numerical integration, indicating with F_t the time average force during the generic integration time interval Δt, and with F_x space average force in the corresponding space interval Dx, Eqs. (3.150) and (3.151) can be written as

$$\begin{cases} m_2(V_{2f} - V_{2i}) = m_1(V_{1i} - V_{1f}) \\ F_t\Delta t = m_1(V_{1i} - V_{1f}) \\ \frac{1}{2}m_1V_{1i}^2 + \frac{1}{2}m_2V_{2i}^2 = \frac{1}{2}m_1V_{1f}^2 + \frac{1}{2}m_2V_{2f}^2 \pm F_x\Delta x \end{cases} \tag{3.152}$$

where V_{ai} and V_{af} refer to the initial and final velocity values of the vehicle a at the initial and final time of the interval Δt considered during the numerical integration.

For each time interval the unknown quantities are the final speed of the vehicles V_{1f} and V_{2f}, the time average force F_t, the space average force F_x, and the global deformation Dx. Since the equations are only three, it is possible to reduce to three the number of unknowns with the following approximation: the time average force in the step Δt is generally different from the space average force, however, for small increments of force, from F_i to $F_f = F_i + \Delta F$, we can assume $F_x = F_t = F_i + \Delta F/2$.

It is thus possible to integrate the equations above by imposing small force increments ΔF rather than of time, obtaining

$$\begin{cases} m_2(V_{2f} - V_{2i}) = m_1(V_{1i} - V_{1f}) \\ \left(F_i + \dfrac{\Delta F}{2}\right)\Delta t = m_1(V_{1i} - V_{1f}) \\ \dfrac{1}{2}m_1 V_{1i}^2 + \dfrac{1}{2}m_2 V_{2i}^2 = \dfrac{1}{2}m_1 V_{1f}^2 + \dfrac{1}{2}m_2 V_{2f}^2 \pm \left(F_i + \dfrac{\Delta F}{2}\right)\Delta x \end{cases} \tag{3.153}$$

The unknowns are, therefore, only the final speed V_{1f} and V_{2f} and the time step Dt. The Δx value is calculated from the curves $F(x)$ of the vehicles, by summing the values of the two deformation increments obtained by each curve, for each increase in force.

The written equations allow a complete forward analysis of the impact and, having the curves $F(x)$ in digitized form, can be implemented in a computer for the step by step integration. The process starts with a force value $F_i = 0$, and initial values of the speeds of the vehicles. At each integration step the final speed values and the consequent time step are calculated. In the next step the force is increased, and the final speed values, previously calculated, become the new initial values.

In the compression phase a positive sign for the energy absorbed is initially used, in the third equation of Eq. (3.153). The process continues until the two vehicles reach the same speed ($V_{2f} \geq V_{1f}$); at this point the vehicles have reached a common speed, and the compression phase is finished.

The force value reached represents the force exchanged by the two vehicles at the maximum deformation instant, and the sum of all the time steps provides the overall compression phase time.

At this point, we can integrate the equations for the restitution phase, decreasing the force at every step and using the negative sign for the energy recovered, in the third equation of Eq. (3.153).

Concerning the curve $F(x)$ to be used in the phase of restitution, a clarification is needed (Cipriani et al., 2002). The portion of the curve $F(x)$ relative to the compression phase is a characteristic of the vehicle, independent of the impact speed, in the field of the typical deformation speed of collisions between vehicles (see Chapter 1: Structural behavior of the vehicle during the impact). By providing the curves $F(x)$ obtained by a test against barrier, for example, to an impact speed of 50 km/h, with the step by step integration described above, we can simulate collisions between the vehicles at all speeds, up to a maximum contact force equal to that of the available curves.

The portion of the curve during the restitution, however, changes depending on the level of force reached, or, equivalently, of the impact speed. In fact, different impact speeds are related to different coefficients of restitution, and this corresponds, from Eq. (3.28), to different areas under the curve $F(x)$. For each impact speed, we have different restitution curves and it is not therefore possible to directly obtain this part of curve from the available crash curves that are related to a specific impact speed and to a precise level of force reached. One possibility is to approximate the part of the restitution curve, for example, by adopting the bilinear model of McHenry (Section 3.1.1), or, in an even more straightforward way, through a single straight line.

In both methods, it is necessary to estimate the coefficient of restitution. In the literature, several expressions correlate the coefficient of restitution to the impact speed (ETS), derived from impact against the barrier. The following is an expression that well approximates the experimental results (Antonetti, 1998):

$$\varepsilon = 0.5992e\left(-0.2508 \cdot V + 0.01934 \cdot V^2 - 0.001279 \cdot V^3\right) \tag{3.154}$$

In our case, however, it is necessary to know the coefficient of restitution starting from an impact speed between vehicles, which, in general, produces contact forces different from those in an impact against the barrier, for the same impact speed. A method to obtaining the value of the searched coefficient, starting from the maximum force reached at the end of compression, it is to apply the step by step integration of Eq. (3.153), to an impact against the barrier, for various impact speeds. In this manner, for each impact speed, the corresponding maximum force reached is obtained. From this relationship, it is possible to calculate, for the maximum force reached in the two vehicles collision, the impact speed against the barrier that produces the same maximum force. Then from Eq. (3.154), the value of the coefficient of restitution can be obtained.

Once determined the coefficient of restitution, it is possible to approximate the curve $F(x)$ following the McHenry approach or to approximate the curve by a straight line, starting from the maximum force value reached and the corresponding deformation value. This line should intersect the abscissa axis into a value of the deformation corresponding to the residual deformation, calculable by imposing Eq. (3.28), as graphically illustrated in Fig. 3.26.

The step by step integration, in the restitution phase, goes on until the contact force becomes null; at that point the vehicle speed variation, the maximum acceleration, the duration, and the deformations both for the compression phase and that of restitution have been found.

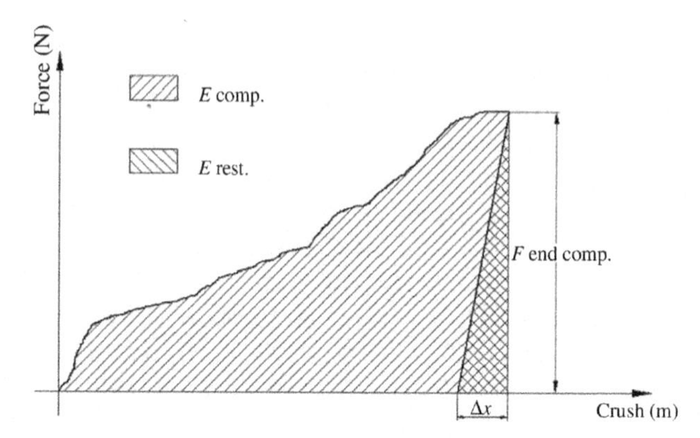

FIGURE 3.26 The areas under the compression and restitution phase of the $F(x)$ curve represent the energy absorbed and recovered, respectively.

In an oblique collision, it is necessary to use a proper $F(x)$ curve and in addition to the speed variations of the centers of gravity of vehicles, there are also the rotational speed variations. In addition to equations (3.153) the two equations (2.21) must then be added, and the impact point (point of application of the resultant force) and the direction of the force (PDOF) have to be defined. In this way, at each integration step, the corresponding center of gravity speed changes and vehicle rotation is obtained.

3.4 Reduced order lumped mass model

In the following a reduced order dynamic models, allowing the analysis and reconstruction of a vehicle's crash, is described. The model consists of a lumped mass model, in which the vehicle discretization affects only its contour. The vehicle is treated in 2D, making the model suitable for crash analysis and reconstruction when 3D phenomena are negligible, for example, no rollover.

The vehicle model considers only rod elements with little possibility to extend or shorten. Rods are without mass and linked together by nodes, as in the FEM. Rotations are allowed only at nodes, and the rods cannot transmit bending moment. The deformation of the vehicle as a whole is the result of nodes displacements only, caused by the impact. Impact forces are transferred to nodes by springs, linked to nodes at one end and a "reference body" at the other end, coinciding with nodal positions before deformation. For the integration of motion equations the inertial properties of the vehicle are applied to the center of gravity of the reference body.

Elastic properties of springs can be determined from load–deformation curves slope obtained from crash tests (Kubiak, 2017; Hormann and Agathos, 2001) or FEM simulations (crash into barrier, front or side crash into a post, etc.).

3.4.1 Model description

Starting from a CAD model of the vehicles, n nodes on the contour are considered, as shown in Fig. 3.27. Typically, 50 nodes are sufficient to describe, in a satisfactory manner, the behavior of a vehicle in the impact.

Inertial properties are attributed to a virtual, nondeformable reference body as the one depicted in Fig. 3.28A. The vehicle profile is linked to the reference body with springs to which vehicles' elastic/plastic properties are assigned at each point, represented by the elastic constants in the two directions defining the reference system. In Fig. 3.28B the anterior part of the left side of the vehicle is shown, in which the springs embodying the system's elastic properties are highlighted.

Forward simulation of the impact is carried out: preimpact known conditions are initially applied to the reference body, that is, position, translational, and rotational velocities. At each time step of the simulation, changes in the parameters are computed and applied as new initial conditions for the successive time step. The reference body translates and rotates as a function of the time step itself.

The simulated impact, at each time step, consists of six different phases: (1) contact identification, (2) PDOF determination, (3) directly connected

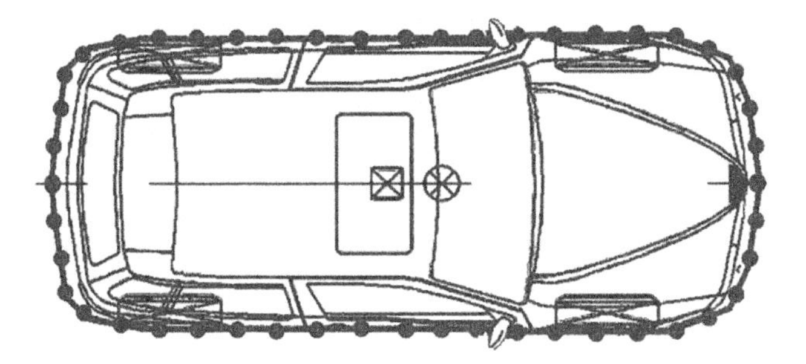

FIGURE 3.27 Discretization of a vehicle contour.

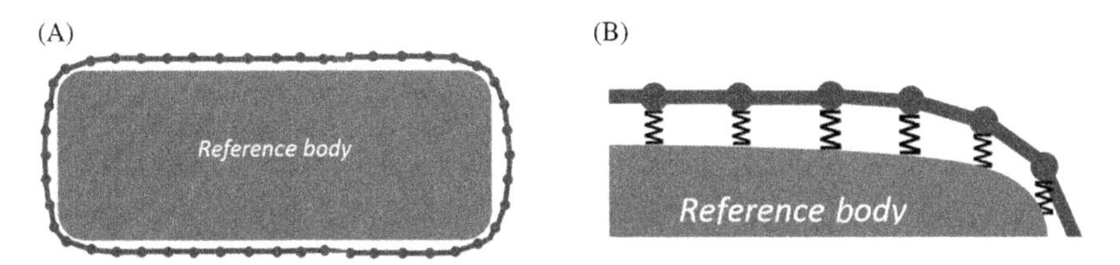

FIGURE 3.28 (A) Representation of the reference body and (B) scheme of connection between reference body and nodes through virtual springs (front part of the left).

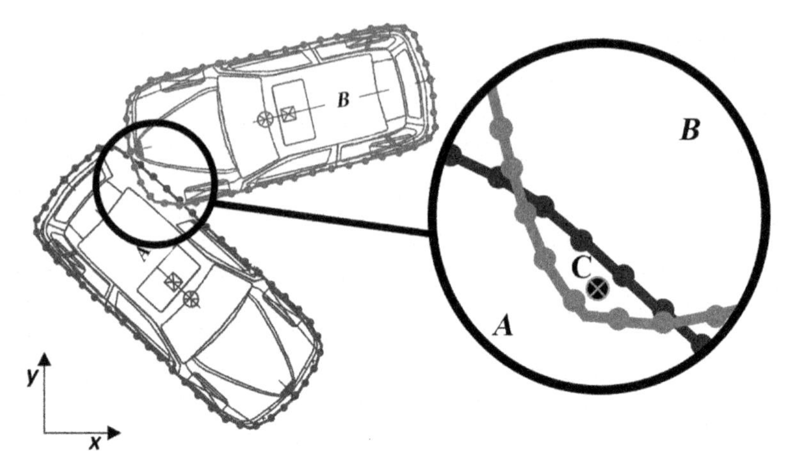

FIGURE 3.29 Crash between two vehicles at a generical time step, with a zoom in on the intrusion area and corresponding center of gravity (impact center).

nodes identification, (4) calculation of forces due to the impact, (5) calculation of final kinematic parameters, and (6) reiteration of the process.

Contact identification

Contact between the vehicles is detected analyzing the position of nodes for both vehicles. At a specified time step, some nodes of a vehicle come into the perimeter of the other, as a consequence of vehicle movement (see Fig. 3.29). The number M_{A-B} of vehicle A nodes positioned inside B is generally different from the number M_{B-A} of vehicle B nodes internal to A; to identify contact, point-in-polygon algorithms can be applied (Hormann and Agathos, 2001). The center of gravity C (center of impact) of the polygon constituted by these nodes—in an amount of $M = M_{A-B} + M_{B-A}$—is then determined; in the zoom of Fig. 3.29, both the common area between the two vehicles (intrusion area) and the corresponding center of impact C are depicted.

Plane of impact and impulse direction

Being the intrusion area generally very small (due to a small time step), the impact plane can be approximated with the line connecting the first two nodes involved in the intrusion, indistinctly for a vehicle or the other. This is represented in Fig. 3.30: the vehicle with the highest number of nodes involved in the intrusion is identified as "master" (B in the figure)—in respect to which all entities of interest are referred, the impact plane among them.

To obtain the impulse direction the empirical relationship between μ and speed ratio S_r, reported in Section 5.1.3, can be used.

Directly connected nodes identification

Once the impulse direction has been determined, it is necessary to identify the nodes of the two vehicles which mutually exchange impact forces. In general terms the number of vehicle A and vehicle B nodes constituting the intrusion area can be different; in this circumstance, a single node of vehicle A can exchange force with multiple nodes of vehicle B or vice versa. Referring to Fig. 3.31, tracing the impulse direction starting from each node of the master vehicle, the nodes of the other vehicle closest to the impulse direction itself are identified. In the case of small intrusion areas the impulse direction can be considered the same for all nodes in the first instance. δ Distances are consequently defined, embodying the intrusion at the considered time step.

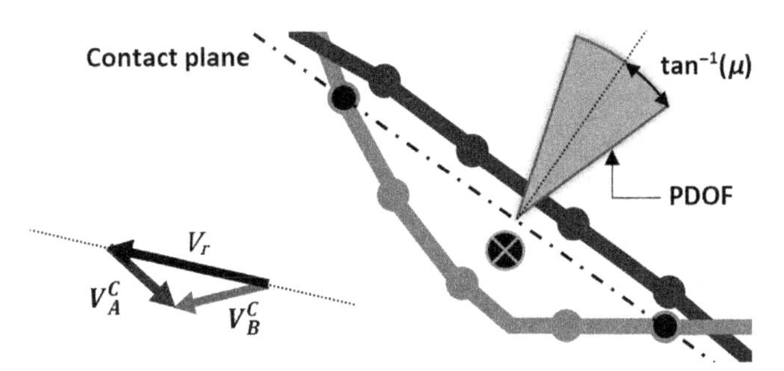

FIGURE 3.30 Identification of impact plane and friction cone; the comparison between V_R direction and friction cone apex angle allows to define the PDOF. *PDOF*, Principal direction of force.

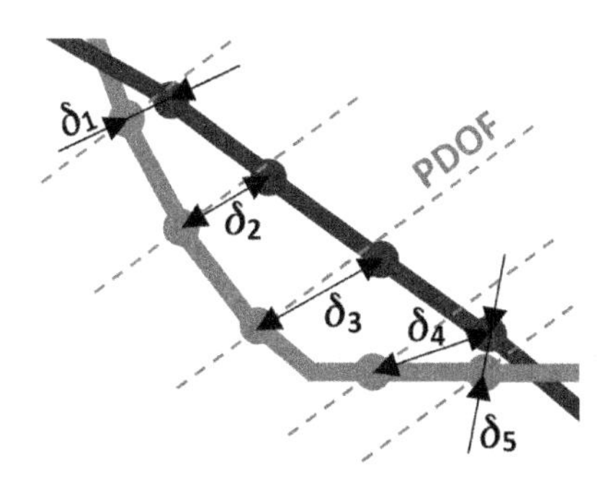

FIGURE 3.31 Identification of directly connected nodes involved in the intrusion.

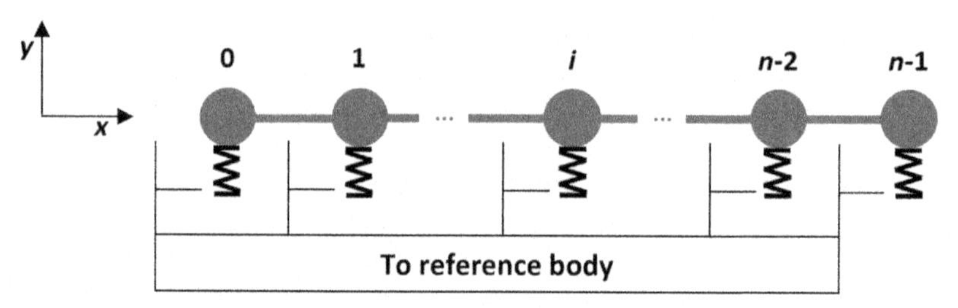

FIGURE 3.32 Visualization of the analyzed vehicle's contour.

Calculation of impact forces

Directly connected nodes mutually exchange impact forces of unknown extent, along the impulse direction. Nodes are connected to the reference body by springs with longitudinal and transversal stiffness, as can be seen in Fig. 3.32 where the rod elements lie on a line for simplicity. The springs behavior can be assumed as linear, as in the Campbell model (see Section 3.1.1) for simplicity, but alternative formulations can be however employed, with or without elastic restitution. Springs stiffness has different values in correspondence of side, front, corner, and wheel nodes. Congruence condition is applied to nodes 0 and $n-1$ (node $0=$ node $n-1$, same x and y coordinates), allowing for a closed geometry of the vehicle.

By indicating with $T_{(i\_i+1)}$ the forces transmitted through the rod linking nodes i and $i+1$, with k_i the elastic constant of the springs relative to node i, with x_i and y_i the node i coordinates, with x_i^0 and y_i^0 the node i coordinates at the beginning time step (nondeformed vehicle) and with F_j the force applied to node j, the following equations can be written:

Equilibrium equations along the x-axis (n equations)

$$
\begin{aligned}
&k_{0x}\left(x_0-x_0^0\right)-T_{(o\_1)x}=0\\
&k_{1x}\left(x_1-x_1^0\right)+T_{(o\_1)x}-T_{(1\_2)x}=0\\
&\ldots\ldots\ldots\ldots\ldots\\
&k_{ix}\left(x_i-x_i^0\right)+T_{(i-1\_i)x}-T_{(i\_i+1)x}=0\\
&\ldots\ldots\ldots\ldots\ldots\\
&k_{jx}\left(x_j-x_j^0\right)+T_{(j-1\_j)x}-T_{(j\_j+1)x}+F_{jx}=0\\
&\ldots\ldots\ldots\ldots\ldots\\
&k_{(n-2)x}\left(x_{n-2}-x_{n-2}^0\right)+T_{(n-3\_n-2)x}-T_{(n-2\_n-1)x}=0\\
&k_{(n-1)x}\left(x_{n-1}-x_{n-1}^0\right)+T_{(n-2\_n-1)x}=0
\end{aligned}
\tag{3.155}
$$

Equilibrium equations along the y-axis (n equations)

$$
\begin{aligned}
& k_{0y}\left(y_0 - y_0^0\right) - T_{(o\_1)y} = 0 \\
& k_{1y}\left(y_1 - y_1^0\right) + T_{(o\_1)y} - T_{(1\_2)y} = 0 \\
& \ldots\ldots\ldots\ldots\ldots \\
& k_{iy}\left(y_i - y_i^0\right) + T_{(i-1\_i)y} - T_{(i\_i+1)y} = 0 \\
& \ldots\ldots\ldots\ldots\ldots \\
& k_{jy}\left(y_j - y_j^0\right) + T_{(j-1\_j)y} - T_{(j\_j+1)y} + F_{jy} = 0 \\
& \ldots\ldots\ldots\ldots\ldots \\
& k_{(n-2)y}\left(y_{n-2} - y_{n-2}^0\right) + T_{(n-3\_n-2)y} - T_{(n-2\_n-1)y} = 0 \\
& k_{(n-1)y}\left(y_{n-1} - y_{n-1}^0\right) + T_{(n-2\_n-1)y} = 0
\end{aligned}
\tag{3.156}
$$

The constancy of distance d_i d_i among consecutive nodes within tolerance ε ($n - 1$ equations)

$$
\begin{aligned}
& (x_1 - x_0)^2 + (y_1 - y_0)^2 < (d_0 + \varepsilon)^2 \\
& (x_2 - x_1)^2 + (y_2 - y_1)^2 < (d_1 + \varepsilon)^2 \\
& \ldots\ldots\ldots\ldots\ldots \\
& (x_i - x_{i-1})^2 + (y_i - y_{i-1})^2 < (d_{i-1} + \varepsilon)^2 \\
& \ldots\ldots\ldots\ldots\ldots \\
& (x_j - x_{j-1})^2 + (y_j - y_{j-1})^2 < (d_{j-1} + \varepsilon)^2 \\
& \ldots\ldots\ldots\ldots\ldots \\
& (x_{n-2} - x_{n-3})^2 + (y_{n-2} - y_{n-3})^2 < (d_{n-3} + \varepsilon)^2 \\
& (x_{n-1} - x_{n-2})^2 + (y_{n-1} - y_{n-2})^2 < (d_{n-2} + \varepsilon)^2
\end{aligned}
\tag{3.157}
$$

T_{i_j} rod internal forces aligned with the axis of the element itself ($n - 1$ equations)

$$
\begin{aligned}
& \frac{T_{(0\_1)x}}{T_{(0\_1)y}} = \frac{x_1 - x_0}{y_1 - y_0} \\
& \frac{T_{(1\_2)x}}{T_{(1\_2)y}} = \frac{x_2 - x_1}{y_2 - y_1} \\
& \ldots\ldots\ldots\ldots\ldots \\
& \frac{T_{(i-1\_i)x}}{T_{(i-1\_i)y}} = \frac{x_i - x_{i-1}}{y_i - y_{i-1}} \\
& \ldots\ldots\ldots\ldots\ldots \\
& \frac{T_{(j-1\_j)x}}{T_{(j-1\_j)y}} = \frac{x_j - x_{j-1}}{y_j - y_{j-1}} \\
& \ldots\ldots\ldots\ldots\ldots \\
& \frac{T_{(n-3\_n-2)x}}{T_{(n-3\_n-2)y}} = \frac{x_{n-2} - x_{n-3}}{y_{n-2} - y_{n-3}} \\
& \frac{T_{(n-2\_n-1)x}}{T_{(n-2\_n-1)y}} = \frac{x_{n-1} - x_{n-2}}{y_{n-1} - y_{n-2}}
\end{aligned}
\tag{3.158}
$$

The unknowns are the x and y nodal coordinates ($2n$) and the T forces components ($2n-2$), resulting in a total amount of $4n-2$ equations in $4n-2$ unknowns. The equations are not linear and must be solved with numerical methods or algorithms such as Newton's or Newton–Raphson's ones. Attempt forces' x and y components must be such as to verify the alignment to the PDOF, that is, $F_{jy}/F_{jx} = \tan^{-1}\text{PDOF}$.

The application of attempt forces is iterated for all nodes of the two vehicles that constitute the intrusion area. Infinite combinations of forces exchanged between nodes exist that allow solving the dynamic problem; however, only one combination is characterized by congruence between intrusion area (represented by a number M of δ distances) and nodal displacements of both vehicles A and B, or else

$$\begin{aligned}\sum_{i=1}^{M}(\delta_i)_x - \varepsilon_\delta \leq \sum_{m=1}^{M_{A\_B}}\left(x_m - x_m^0\right) + \sum_{p=1}^{M_{B\_A}}\left(x_p - x_p^0\right) \leq \sum_{l=1}^{M}(\delta_l)_x - \varepsilon_\delta \\ \sum_{i=1}^{M}(\delta_i y_x) - \varepsilon_\delta \leq \sum_{m=1}^{M_{A\_B}}\left(y_m - y_m^0\right) + \sum_{p=1}^{M_{B\_A}}\left(y_p - y_p^0\right) \leq \sum_{l=1}^{M}(\delta_l)_y - \varepsilon_\delta\end{aligned} \tag{3.159}$$

where ε_δ is a tolerance for the intrusion area, while $(\delta)_x$ $(\delta_l)_x$ and $(\delta)_y$ $(\delta_l)_y$ are the x and y components of the ith distance δ. If the conditions expressed in Eq. (3.159) are not satisfied, the attempt values of x and y components of forces between the mth node (for vehicle A) and pth node (for vehicle B) are modified at each iteration, adding a value equal to

$$\begin{aligned}(\Delta F_{mp})_x = b\left[(\delta_l)_x - \left(x_m - x_m^0\right) - \left(x_p - x_p^0\right)\right] \\ (\Delta F_{mp})_y = b\left[(\delta_l)_y - \left(y_m - y_m^0\right) - \left(y_p - y_p^0\right)\right]\end{aligned} \tag{3.160}$$

with b representing a multiplication factor. The process is repeated until Eq. (3.159) is fulfilled; the solution is quicker based on the selected value of b.

Calculation of final kinematic parameters

Once the nodal forces are determined through iterations, it is possible to apply the laws of rigid body dynamics to the reference body. The center of gravity of vehicle a undergoes a global change in translational ΔV and rotational $\Delta\omega$ velocities ruled by the two impulse theorems, or else:

$$\begin{aligned}\Delta V_a = \left(R_a + \sum_{q=1}^{4}(F_b)_q\right)\frac{\Delta t}{m_a} \\ \Delta\omega_a = \frac{1}{J_a}\left(R_a \wedge h_a + \Delta t\sum_{q=1}^{4}(F_b)_q \wedge (h_b)_q\right)\end{aligned} \tag{3.161}$$

where R_a is the vector sum of nodal forces applied to point C, F_b the braking actions applied to the single wheel, h_b the corresponding arm, m_a and J_a the total mass and moment of inertia of the ath vehicle and Δt the time step. Forces F_b exchanged between the wheels, and the road surface can be computed by the adherence circle (or ellipse) model (Brach and Brach, 2005); more sophisticated models can, however, be referred to Li et al. (2013), also allowing to select an appropriate road-tire coefficient of friction for the specific event (Vangi and Virga, 2007).

After the two kinematic parameters in Eq. (3.161) have been obtained for each vehicle, it is possible to apply them to the corresponding reference bodies and reiterate the steps previously described until the two bodies disengage (end of the impact phase) or stop in the rest position.

The calculation procedure applies similarly in the compression and the restitution phases, the latter occurring when the direction of vehicles' superimposition inverts (McHenry and McHenry, 1997; Goldsmith, 2001): different elastic constants k must be applied, according to the desired load-crush law of the vehicle.

References

Antonetti, V.W., 1998. Estimating the coefficient of restitution of vehicle-to-vehicle bumper impacts. In: SAE Paper 980552. Amatech Review.

Brach, R.M., Brach, M.R., 2005. Vehicle Accident Analysis and Reconstruction Methods. In: SAE International, Warrendale, PA. ISBN 0-7680-0776-3.

Campbell, K.E., 1974. Energy basis for collision severity. In: SAE Paper 740565. Society of Automotive Engineers, Inc., Warrendale, PA.

Cipriani, A.L., Bayan, F.P., Woodhouse, M.L., Cornetto, A.D., Dalton, A.P., Tanner, C.B., Timbario, T.A., Deyerl, E.S., 2002, Low-speed collinear impact severity: a comparison between full-scale testing and analytical prediction tools with restitution analysis. In: SAE Paper 2002-01-0540. FTI/SEA Consulting, Quan, Smith & Associates.

Goldsmith, W., 2001. Impact—The Theory and Physical Behaviour of Colliding Solids. Dover Publications, Inc, Mineola, NY.

Hormann, K., Agathos, A., 2001. The point in polygon problem for arbitrary polygons. Comput. Geom. 20, 131–144.

Huang, M., 2002. Vehicle Crash Mechanics, first ed. CRC Press, Boca Raton, FL.

Kubiak, P., 2017. Nonlinear approximation method of vehicle velocity V_t and statistical population of experimental cases. Forensic Sci. Int. 281 (2017), 147–151.

Li, Y., Sun, W., Huang, J., Zheng, L., 2013. Effect of vertical and lateral coupling between tyre and road on vehicle rollover. Veh. Syst. Dyn. 51 (8), 1216–1241.

Macmillan, 1983. Dynamics of vehicle collisions. In: Proceedings of the International Association for Vehicle Design, Editor-in-Chief M.A. Dorgham.

McHenry, R.R., McHenry, B.G., 1997. Effects of restitution in the application of crush coefficients. In: SAE Paper No. 970960. Society of Automotive Engineers, Inc., Warrendale, PA.

Neptune, J.A., 1998. Crush stiffness coefficients, restitution constants, and a revision of CRASH3 and SMAC. In: SAE Paper 980029. Neptune Engineering, Inc.

Siddal, D.E., Day, T.D., 1996. Updating the vehicle class categories. In: SAE Paper 960897. Engineering Dynamics Corp.

Varat, M.S., Husher, S.E., 2000. Vehicle impact response analysis through the use of acceleration data. In: SAE Paper 2000-01-0850. KEVA Engineering.

Varat, M.S., Husher, S.E., 2003. Crash pulse modeling for vehicle safety research. In: 18th ESV Paper.

Vangi, D., Virga, A., 2007. Evaluation of emergency braking deceleration for accident reconstruction. Veh. Syst. Dyn. 45 (10), 895–910.

Wei, Z., Robbersmyr, K.G., Karimi, H.R., 2017. Data-based modeling and estimation of vehicle crash processes in frontal fixed-barrier crashes. J. Franklin Inst. 354 (12), 4896–4912.

Chapter 4

Energy loss

Chapter Outline

In a generic impact, a part of the initial kinetic energy of the system is converted into vehicle deformation energy. In a crash test or a numerical simulation, the dissipated energy can be computed as a simple difference between the kinetic energies of the vehicles before and after the impact; in a real accident, instead, the deformation energy or energy loss, can only be estimated starting from vehicle deformations. The ex post evaluation of the energy dissipated in vehicles assumes particular importance in the analysis and reconstruction of road accidents (Vangi, 2008; Brach and Brach, 2005).

The application of contact forces causes both direct and induced deformations, and therefore the energy loss is equal to the work of the contact forces themselves. The ex post evaluation of the energy loss can be carried out starting from the evaluation of the work of the contact forces, through the knowledge of the force-deformation law $F(x)$ and the measurement of the deformations. In an alternative way the energy loss can be computed, in a more qualitative way, through a comparison, carried out with a parameter called EES (energy equivalent speed), with similar deformations on crashed vehicles, of which the dissipated energy is known or with mixed methods (triangle method). The various methods most commonly used are illustrated in the following sections.

Vehicle Collision Dynamics. DOI: https://doi.org/10.1016/B978-0-12-812750-6.00004-4

4.1 The classical approach to estimate the energy loss

This procedure allows, using the schematization of the $F(x)$ curve proposed by Campbell (1974) (see Chapter 3: Models for the structural vehicle behavior), to evaluate the energy absorbed in the compression phase. Then considering the restitution coefficient, we can also calculate the energy restored elastically and therefore the entire energy dissipated.

Referring briefly what is stated in Chapter 3, Models for the structural vehicle behavior, the schematization proposed by Campbell schematizes the vehicle as a spring–mass system, wherein the compression part of the $F(x)$ curve is approximated with a straight line. In particular, reference is made not to dynamic deformations but to residual ones, or to deformations that remain visible after the impact. Furthermore, the force is normalized with respect to the width L of the deformed area, which in the case of frontal impacts against the barrier coincides with the width of the front. With this approximation, therefore, the force F per unit area is expressed as

$$F = BC + A \tag{4.1}$$

where C is the residual deformation (crush), A the intercept, and B the slope (Nystrom et al., 1991; Siddal and Day, 1996).

G indicates the part of energy due to elastic deformations, below which there is no residual deformation.

Since the coefficients A and B are related to the unit of width of the deformed zone, the global kinetic energy absorbed during the deformation is equal to

$$E_a = \int_0^L \left[G + \int_0^C F(C)dC \right] dl = \int_0^L \left(G + AC + \frac{BC^2}{2} \right) dl \tag{4.2}$$

In evaluating the absorbed energy the hypothesis is that the deformation of the vehicle is uniform in a vertical plane, and the extension and depth of the deformation are completely characterized by the profile of the vehicle projected on a plane parallel to the road, as exemplified in Fig. 4.1. To define this profile, in practice, it is sufficient to have a number n of deformation depth measurements (Tumbas and Smith, 1988; Prasad, 1990; Prasad, 1991).

The deformation depth C is linearized between two adjacent measurement points, and the entire profile of the deformed area is approximated with a broken line (see Fig. 4.2). With this approximation the integral of Eq. (4.2) can be evaluated through the sum of $n - 1$ integral with a linear trend of C:

$$E_a = \sum_1^{n-1} \int_0^l \left(G + AC + \frac{BC^2}{2} \right) ds \tag{4.3}$$

where s indicates the coordinate between one measurement point and another, as shown in Fig. 4.3, and

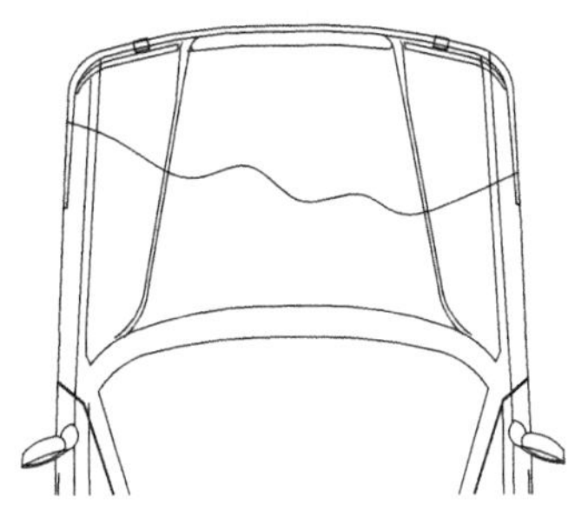

FIGURE 4.1 Vehicle's deformation scheme, with deformation projected on a plane parallel to the road.

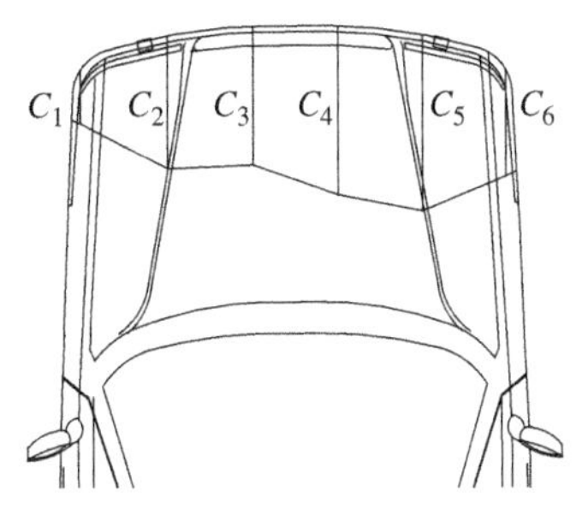

FIGURE 4.2 Scheme of measurement of deformation depths. Approximation of the deformed profile with a broken line.

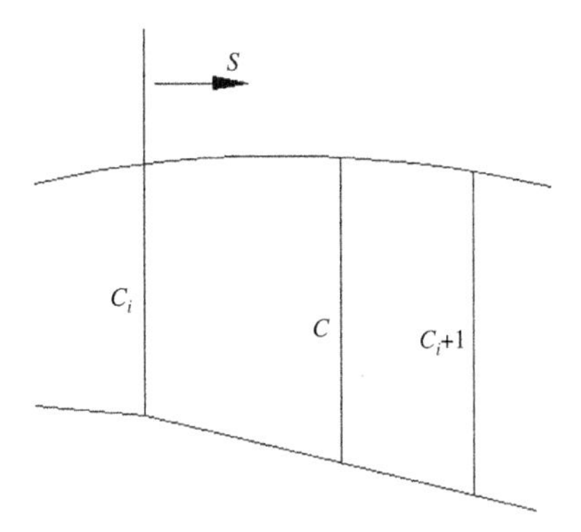

FIGURE 4.3 Coordinate s between two successive deformation measurements.

$$l = \frac{L}{n-1} \tag{4.4}$$

Since between point ith and the next, we have

$$C = C_i + (C_{i+1} - C_i)\frac{s}{l} \tag{4.5}$$

we can write

$$ds = \frac{l}{C_{i+1} - C_i} dC \quad \text{con } 0 \leq s \leq l \text{ e } C_i \leq C \leq C_{i+1} \tag{4.6}$$

and the single term of the summation (4.3) becomes

$$E_{ai} = \int_{C_i}^{C_{i+1}} \left(G + AC + \frac{BC^2}{2} \right) \frac{l}{(C_{i+1} - C_i)} dc$$
$$= l \left[G + \frac{A}{2}(C_{i+1} + C_i) + \frac{B}{6}\left(C_{i+1}^2 + C_{i+1}C_i + C_i^2\right) \right] \tag{4.7}$$

For a generic number n of measures the summation can, therefore, be written as

$$E_a = \frac{L}{n-1}\left[(n-1)G + \frac{A}{2}\left(C_1 + \sum_2^{n-1} 2C_i + C_n \right) + \frac{B}{6}\left(C_1^2 + \sum_2^{n-1} 2C_1^2 + C_n^2 + \sum_1^{n-1} C_i C_{i+1} \right) \right] \tag{4.8}$$

With four points, for example, we have

$$E_a = \frac{L}{6}\left[6G + A(C_1 + 2C_2 + 2C_3 + C_4) + \frac{B}{3}\left(C_1^2 + 2C_2^2 + 2C_3^2 + C_4^2 + C_1C_2 + C_2C_3 + C_3C_4\right) \right] \tag{4.9}$$

Eq. (4.8) allows calculating the energy absorbed in the compression phase, starting from the n crush C measures and the values of the coefficients A and B, available for various vehicles or tabulated by vehicle classes (https://www.nhtsa.gov).

For medium- or high-speed impacts, with relative impact speeds more than 30 km/h, the energy elastically recovered is negligible compared to the absorbed percentage, and therefore the deformation energy can be well approximated with the absorbed one.

To evaluate the energy recovered, it is possible to use the restitution coefficient, and from Eq. (3.28) the deformation energy can be obtained as follows:

$$E_d = E_a - E_r = E_a(1 - \varepsilon^2) \tag{4.10}$$

The method described for determining the deformation energy is based on the use of the stiffness coefficients of vehicles A and B, which are derived from crash tests against a rigid flat barrier. The coefficients approximate the relationship between force and deformation in a sufficiently reliable manner only when the deformation is comparable to the one which occurs in the collision against the barrier, with almost uniform depth in the vertical direction. Furthermore, the coefficients are relative to the normalized force, that is, it is assumed that the vehicle stiffness is uniform on the damaged area, both vertically and longitudinally.

The limits of the method are derived from these assumptions; the method is all the more reliable, the more the damage to the vehicle involved in a real accident is uniform in the vertical plane. In the case of very irregular or extensive deformations the method is not applicable, since the coefficients A and B are not representative of the stiffness of the vehicle for these types of deformations. Examples are when a vehicle is wedged under a heavy vehicle (the upper part is deformed much more than the lower one), in case of roll-over (damage not due to a collision with a barrier) or the event of sliding impact between vehicles (high tangential component of the forces referring to the damaged surfaces).

The method does not apply even in the case of heavy industrial and two-wheeled vehicles, due to the lack of stiffness data relating to these vehicles.

4.2 Correction for oblique impacts

The energy thus calculated assumes that the deformation depths are measured parallel to or perpendicular to the vehicle's axis, depending on whether the deformations are on the front/rear or side. In fact, in the energy calculation the contact force is considered applied in these directions, since these are the conditions of the crash tests in which the $F(x)$ curves and the coefficients A and B are obtained.

In actual impacts, however, often the direction of the resulting forces is not aligned along the longitudinal or perpendicular direction to the vehicle axis; this implies that the $F(x)$ curves from which to derive the coefficients A and B are not those obtained in frontal or lateral impacts but those obtained from angled impacts, as in the real impact. Moreover, the deformations depths should be measured along the direction in which the forces are applied.

However, this is not very easy in practice, due to the difficulty of establishing the direction of the forces and the lack of experimental curves in various directions. It is, therefore, preferable to measure the deformations in a longitudinal or perpendicular direction, calculate the energy absorbed with Eq. (4.8), using the coefficients A and B relative to frontal or lateral impact directions, and multiply the result by a corrective factor f that takes into account the inclination α of the force with respect to the vehicle axis [principal direction of the force (PDOF)], as shown in Fig. 4.4.

However, it must be highlighted that the method of evaluating the dissipated energy starting from the measurement of the deformations is applicable only in the case of impacts in which the forces are not too inclined with respect to the surface of each vehicle. In fact, in the case of important tangential components of force the assumed force–deformation curves, obtained in impacts with forces in a direction normal to the surfaces of the vehicle, are not yet valid. This excludes from this type of methodology, the

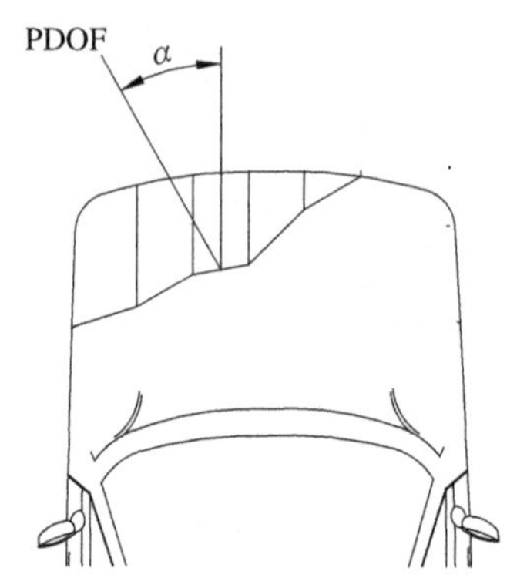

FIGURE 4.4 Contact force inclined by an angle α with respect to the vehicle axis.

collisions where the vehicles slide over one another without reaching a common speed on the contact plane (sideswipe).

In technical literature (Fonda, 1999; Smith and Noga, 1982; Woolley et al., 1985; McHenry, 2001; Vangi., 2009b), various corrective factors have been proposed, some of which are reported as it is still used in some commercial software. Historically, the first corrective factor used, introduced by McHenry (CR81, 1981; Smith and Noga, 1982; Woolley et al., 1985), was

$$f = \frac{1}{\cos^2\alpha} = 1 + \tan^2\alpha \tag{4.11}$$

This corrective factor, used in software, such as CRASH3 and EDCRASH (CR81, 1981; Woolley et al., 1985, McHenry and McHenry, 1986), can be used for inclination α_n up to 45 degrees, where it provides a corrective value of 2. Above this value the correction is always considered equal to 2. The origin of this factor derives from considering that the force per unit of width L of the deformed zone and the depth of the deformation are equal, respectively, to $F_n/\cos\alpha$ and $C_n/\cos\alpha$, respectively, having indicated with F_n and C_n the force and deformation components in the direction along the vehicle axis. The work of the force, per unit of width of the deformed area, is then given by

$$W = \int_0^C F dc = \int_0^C \frac{1}{\cos^2\alpha} F_n dc_n = \frac{1}{\cos^2\alpha} \int_0^C F_n dc_n \tag{4.12}$$

Subsequently, the hypotheses were refined, and new corrections were made. Considering that the maximum force does not manifest itself in the direction of the maximum deformation, as there are also friction forces between the sheets, with a friction coefficient μ, a new correction (McHenry, 2001) is obtained:

$$f = 1 + \mu \tan\alpha \tag{4.13}$$

This corrective factor, used for example in the CRASH4 version, is valid up to 60 degrees angles and requires knowledge or estimation of the friction coefficient between the sheets in contact, which is not always easy to evaluate. The typical values of this friction coefficient are between 0.45 and 0.55.

A corrective factor (Fonda, 1999), which gives similar results to Eq. (4.13), derives from the same considerations made to derive Eq. (4.11), but it considers that the width L of the deformed area, in a direction perpendicular to the direction of the force, is $L\cos\alpha$ and therefore, by multiplying the force $F_n/\cos\alpha$ per unit of width by the width of the deformed area, the factor $\cos\alpha$ is eliminated, thus obtaining

$$f = \frac{1}{\cos\alpha} \tag{4.14}$$

This corrective factor is also usable up to 60 degrees, beyond which it is set at 2.

Eq. (4.14) represents a good compromise between precision and ease of use, requiring only PDOF as input data.

A more general corrective factor can be derived from the following considerations (Vangi., 2009b), introducing the direction of the resulting deformation [principal direction of the deformation (PDOD)].

Consider the general case of collision, with a generic vehicle configuration, as shown in Fig. 4.5.

The impulse acting on the vehicle 1 can be decomposed into normal and tangential components relative to the impact plane (see Fig. 4.6).

The work done by the resultant of the contact forces, equal to the deformation energy, can be expressed through the mean value theorem as

$$L = E_d = \int_0^s F(s)ds = F\Delta s = F\Delta t\dot{s} = I\dot{s} \tag{4.15}$$

where F and $\dot{s}$ are the mean force and the mean deformation speed during the impact. Therefore for the two components of the impulse acting on each vehicle, it is possible to write

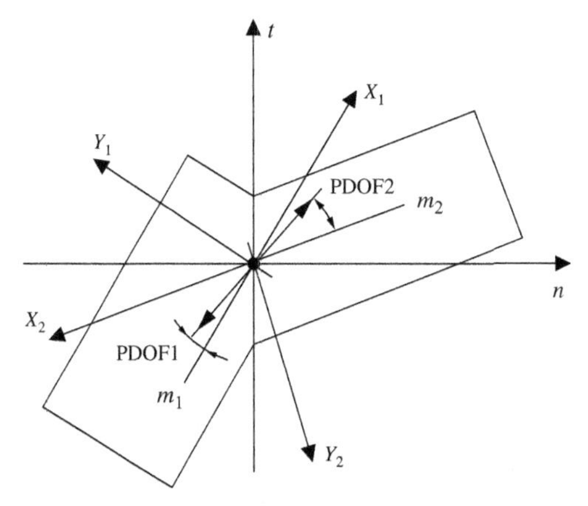

FIGURE 4.5 Reference system and principal directions of forces.

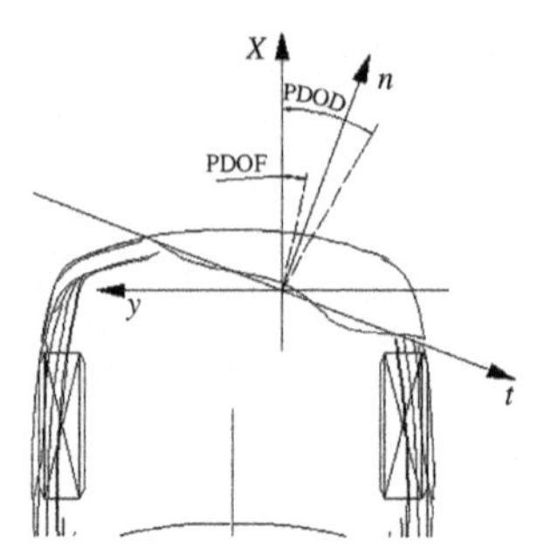

FIGURE 4.6 Principal direction of force and deformation.

$$\begin{aligned} E_{dX} &= I_X \dot{s}_X \\ E_{dY} &= I_Y \dot{s}_Y \end{aligned} \tag{4.16}$$

The global energy is thus

$$E_d = E_{dX} + E_{dY} = I_X(\dot{s}_X + \mu \dot{s}_Y) \tag{4.17}$$

where $\mu = \tan(\text{PDOF})$ and can be rewritten as

$$E_d = E_{dX}\left[1 + \mu \frac{\dot{s}_Y}{\dot{s}_X}\right] \tag{4.18}$$

Hypothesizing that the crush in the X and Y directions ceases at the same instant, it can be stated that

$$\frac{\dot{s}_Y}{\dot{s}_X} = \frac{s_Y}{s_X} = tg(\text{PDOD}) \tag{4.19}$$

Eq. (4.18) thus changes to

$$E_d = E_{dX}[1 + tg(\text{PDOF})tg(\text{PDOD})] \tag{4.20}$$

The overall energy dissipated in the impact is given by the sum of the energy dissipated on each vehicle, and therefore

$$\begin{aligned} E_d = E_{d1} + E_{d2} = E_{dn1}&[1 + tg(\text{PDOF}_1)tg(\text{PDOD}_1)] \\ + E_{dn2}&[1 + tg(\text{PDOF}_2)tg(\text{PDOD}_2)] \end{aligned} \tag{4.21}$$

Eq. (4.21) is true if we consider any surface of the vehicle, not just the front surface, so that the meaning of the correction found is quite general.

In this approach the PDOD was introduced, which is conceptually different from the main direction of the force (PDOF). Deformation is a quantity that can be measured directly, unlike force and more easily identifiable starting from the deformed shape of the vehicle. With the same impact configuration (arrival directions of the vehicles, masses, impact center, etc.) and of the PDOFs, vehicles with different structures can show different principal directions of deformation.

Generally, however, the deformation direction in the direct damage area has a direction practically coinciding with that of the resultant of forces, so that the PDOD measurement can reliably estimate the PDOF.

Considering PDOF = PDOD, the correction to be applied to normal energy to obtain the global dissipated energy is formally identical to the first correction proposed by McHenry in 1976, equal to $(1 + \tan^2\text{PDOD})$. The substantial difference between the two corrections lies in the measurement of deformation depths. In McHenry's work the correction applies to the energy calculated starting from the deformation values measured in the normal direction, for example, at the front of the vehicle. In this case, instead, the correction is applied to the work done by the normal component of the resultant force that, as analyzed next, while being equal to the energy dissipated by normal forces, does not coincide with the first-mentioned energy.

4.2.1 Measurement of deformation depths

Applying the method of calculating the deformation energy based on the PDOD, the following procedure is applied for the measurement of deformations (Vangi, 2009b) (see Fig. 4.7):

1. Select some measurement points (measurement stations) on the vehicle's undeformed profile.
2. The damaged profile of the vehicle is detected. The survey can be carried out manually, with the classical method of measuring deformations, or with any other automatic method, such, for example, with photogrammetric or laser scanner techniques. On this deformed profile the position of the previously selected measurement points (homologous points) can be identified.
3. The undeformed and deformed profiles are superimposed, and the homologous points are connected to determine the direction of the PDOD of

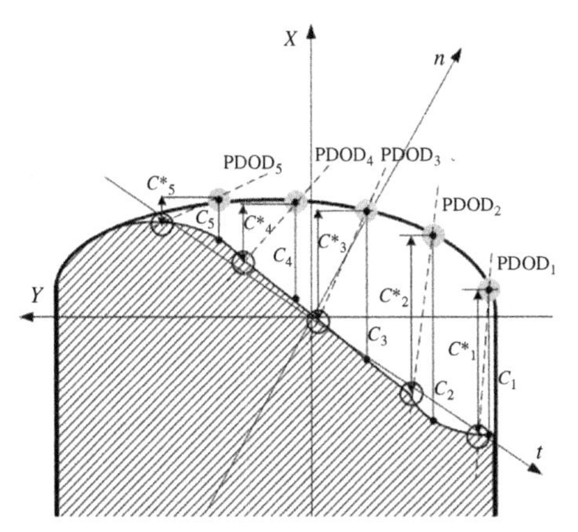

FIGURE 4.7 Deformations depths measurement with the method based on PDOD. *PDOD*, Principal direction of the deformation.

each point (angle between the segment joining the measurement points and the vehicle axis).

4. The distance between the homologous points in the deformed and undeformed profile along the direction of the vehicle axis is projected to determine the residual C_i^* deformation depth.

Each point, selected on the undeformed profile, during the impact moves to the position identified on the deformed profile. Different angles from point to point can exist since the vehicle structure is not homogeneous and because the applied forces are not constant, especially in correspondence of the areas of direct damage with those of indirect damage. The correction factor for the calculation of the deformation energy must, therefore, be applied for each zone (for each measurement station) in which the PDOD can be considered constant.

Fig. 4.7 shows the procedure for measuring the residual deformations C_i^* along the X direction. The deformation energy can then be calculated directly by applying the following expression:

$$E_d = \frac{W}{k-1}\sum_{1}^{k-1}\left\{\left[G + \frac{A}{2}\left(C_i^* + C_{i+1}^*\right) + \frac{B}{6}\left(C_i^{*2} + C_i^* C_{i+1}^* + C_{i+1}^{*2}\right)\right]\left[1 + \tan \mathrm{PDOF} \tan\left(\frac{\mathrm{PDOD}_i + \mathrm{PDOD}_{i+1}}{2}\right)\right]\right\} \tag{4.22}$$

where the correction factor is applied to each term of the summation, instead of to the entire energy in the X direction.

From Fig. 4.7, it is observed that the PDOD influences not only the corrective factor but also the specific terms C_i^* and therefore also the energy in the X direction. Therefore the illustrated procedure does not imply a simple corrective factor but a complete reformulation of the energy calculation of deformation and allows a more accurate estimate [on average with errors around 10% against 20% of the classical method (Vangi, 2009b); the latter seems to produce an overestimation of the deformation energy].

4.3 Determination of *A* and *B* stiffness coefficients from the residual crush

The coefficients A and B are available in the literature for a wide class of vehicles or are defined as an average value for each given class of vehicles, identified according to the wheelbase or mass (https://www.nhtsa.gov). It should be emphasized, however, that since this approach was developed above all in the United States, the data were taken from the American vehicle fleet, which may have stiffnesses and in some cases different from European vehicles. Furthermore, each class includes many vehicles of different characteristics, and there is not always a correspondence between vehicle and stiffness coefficients. It is, therefore, preferable to determine the stiffness coefficients for the specific vehicle; this can be done starting from data of a crash test against undeformable barrier (Neptune et al, 1992).

4.3.1 Crash against fix rigid barrier

Consider a frontal crash test on a vehicle of mass m against a rigid undeformable barrier, at speed V. From the measurement of the deformation depths C_i, we can calculate the energy absorbed in the compression phase with Eq. (4.8) and equalize it to the initial kinetic energy:

$$E_a = \frac{L}{n-1}\left[(n-1)G + \frac{A}{2}\left(C_1 + \sum_2^{n-1} 2C_i + C_n\right) + \frac{B}{6}\left(C_1^2 + \sum_2^{n-1} 2C_1^2 + C_n^2 + \sum_1^{n-1} C_i C_{i+1}\right)\right] = \frac{1}{2}mV^2 \tag{4.23}$$

However, in this way, we obtain a single equation with the two unknowns A and B; it is, therefore, convenient to express the coefficients A and B as a function of the coefficients b_0 and b_1, using Eq. (3.15).

The coefficients b_0 and b_1 correlate the speed of impact with the crushing C of the vehicle, according to the Campbell mass–spring model, see Eq. (3.11). The coefficient b_0 represents the maximum speed below which there are no permanent deformations. Coefficient b_0 is practically constant for all vehicles, with a value close to 8 km/h = 2.22 m/s for the front and rear impacts. This assumption is justified by the fact that the minimum speed so that no permanent deformations occur in the structure is subject to precise regulations, which must be complied with by all vehicles.

The following expression is obtained:

$$V^2 = \frac{1}{n-1}\left[(n-1)b_0^2 + b_0 b_1\left(C_1 + \sum_2^{n-1} 2C_i + C_n\right) + \frac{b_1^2}{3}\left(C_1^2 + \sum_2^{n-1} 2C_1^2 + C_n^2 + \sum_1^{n-1} C_i C_{i+1}\right)\right]^2 \tag{4.24}$$

from which, known b_0, it is possible to obtain b_1

$$b_1 = \frac{-b_0\rho + \sqrt{(b_0\rho)^2 - 20\delta(b_0^2 - V^2)/3}}{2\delta/3} \tag{4.25}$$

with

$$\begin{aligned} \rho &= C_1 + 2(C_2 + C_3 + C_4 + C_5) + C_6 \\ \delta &= C_1^2 + 2(C_2^2 + C_3^2 + C_4^2 + C_5^2) + C_6^2 + C_1C_2 + C_2C_3 + C_3C_4 + C_4C_5 + C_5C_6 \end{aligned} \tag{4.26}$$

The coefficient b_0 can be set equal to 2.22 m/s or obtained with the following energy considerations. The elastic energy G, or the energy corresponding to the initial part of the impact in which residual deformations are not yet generated, is given by (see Section 3.1.1)

$$G = \frac{A^2}{2B} \tag{4.27}$$

It can be assumed, as a first approximation, that the energy elastically recovered is equal to G, that is,

$$G = \frac{1}{2} m \overline{V}^2 \tag{4.28}$$

from which, substituting Eq. (4.27) into Eq. (4.28), we get the A value:

$$A = \overline{V}\sqrt{mB} = \varepsilon V \sqrt{mB} \tag{4.29}$$

From Eqs. (3.15) and (4.29), we can get the b_0 value:

$$b_0 = \overline{V}\sqrt{L} = \varepsilon \mathrm{V} \sqrt{L} \tag{4.30}$$

At this point, known b_0 and obtained b_1, we can determine from Eq. (3.15) the searched coefficients A and B.

If we can consider the C_i coefficients having the same values, Eq. (4.25) is simplified in

$$b_1 = \frac{V - b_0}{C} \tag{4.31}$$

In general, if the deformation profile produces C_i coefficients with similar values, with a good approximation, we can consider a mean C coefficient[1] and apply Eq. (4.31) without making a big mistake.

4.3.2 Crash against a mobile rigid barrier

In the event of a side or rear impacts, in which the crash test is carried out with a nondeformable mobile barrier that impacts the vehicle, the kinetic energy of the barrier must also be taken into account, before and after the impact. The energy balance can be written as

$$\frac{1}{2} m_v V_v^2 + \frac{1}{2} m_b V_b^2 = \frac{1}{2} m_v \overline{V}_v^2 + \frac{1}{2} m_b \overline{V}_b^2 + E_d \tag{4.32}$$

where the subscript v indicates the quantities related to the vehicle and the subscript b those relative to the barrier; the quantity over signed refers to the end of the impact, those not over signed to the initial time.

1. The mean C coefficient can be evaluated as $\rho/10$.

Since the barrier is nondeformable, the kinetic energy is dissipated only in the deformations of the vehicle. The dissipated energy can be written in the following form (see Section 4.5):

$$E_d = \frac{1}{2} m_v \mathrm{EES}^2 \tag{4.33}$$

From Eqs. (4.32) and (4.33), we get

$$\mathrm{EES} = \sqrt{\frac{m_b}{m_v}\left(V_b^2 - \overline{V}_b^2\right) + \left(V_v^2 - \overline{V}_v^2\right)} \tag{4.34}$$

The value of the coefficient b_1 can be determined from Eq. (4.25), using instead of the impact velocity V, the EES:

$$b_1 = \frac{-b_0\rho + \sqrt{(b_0\rho)^2 - 2\theta\delta\left(b_0^2 - \mathrm{EES}^2\right)/3}}{2\delta/3} \tag{4.35}$$

4.3.3 Oblique crash against fix rigid barrier

If a crash test against an inclined, immovable rigid barrier is available, the main direction of the forces, PDOF, is no longer aligned with the vehicle axles and, as seen in Section 4.2, the energy calculated from the deformation measurement must be multiplied by a correction factor. We assume, without losing the generality, the following correction [see Eq. (4.11)]:

$$E_d = \frac{E_a}{\cos^2\theta} \tag{4.36}$$

where E_a is the energy calculated with Eq. (4.23). Finding E_a and setting it in terms of speed, we can write

$$E_a = \frac{1}{2} m_v (\mathrm{EES}\cos\theta)^2 \tag{4.37}$$

where EES, if the final speed is not zero, is given by the energy balance. Taking into account Eq. (4.33): $\mathrm{EES} = \sqrt{V^2 - \overline{V}^2}$, we get

$$V_{\mathrm{eff}} = \mathrm{EES}\cos\theta \tag{4.38}$$

Therefore also in this case, the value of the coefficient b_1 can be determined by Eq. (4.25) using, in place of the speed V, the actual speed V_{eff}.

The calculation of the A and B coefficients with the methods described earlier allows a higher accuracy to be obtained with respect taking the coefficients from the table. It must, however, be observed that also in this way errors may happen due to the uncertainties in the measured values C_i. Moreover, for the same vehicle and impact velocity against the barrier, sometimes different values of the deformation depth are obtained. This is principally due because, although controlled, the crash test is never repeated identical to itself and the vehicle can undergo, for example, rotations and deformations not perfectly perpendicular to its axis, the bumper can partially detach and lead to different measures from case to case. This can also be verified directly by looking at the crash results available on specialized sites.

Consider, for example, the frontal crash tests no. 3127 and 3113 available on the NHTSA database (https://www.nhtsa.gov) conducted on the same type of GM Saturn vehicle at a speed of 48.3 km/h against an immovable rigid barrier. The measurement of the deformation provides the C_i values reported in Table 4.1, which also shows the values of b_1 calculated by Eq. (4.25):

The differences are high and are not justified by the small mass difference, about 60 kg, between the two vehicles (as can easily be verified); observing the photographs of the vehicles after the collision (Fig. 4.8) is noted as in test 3113, the deformation is asymmetric and a little oblique; the inclination of the deformed profile implies a PDOF different from zero and also using Eq. (4.38), a significant difference is obtained between the coefficients b_1. This can partly be attributed to the bumper in test 3113, deformed in such a way as to produce uncertainties in the measurement of C_i, and partly by the elastic instability process governing the deformation, that for small differences in the forces angle can produce different failure mechanisms and consequent differences in the deformation extension. These differences cannot always be compensated with the correction factors referred to in Section 4.2, based on the hypothesis of homogeneity of the structure and globally isotropic behavior.

TABLE 4.1 C_i values from two different NHTSA crash tests on the same vehicle model.

No. of test	3127	3113
c_1	0.415	0.198
c_2	0.480	0.302
c_3	0.506	0.401
c_4	0.512	0.443
c_5	0.472	0.400
c_6	0.423	0.354
b_1	23.4	29.8

FIGURE 4.8 Vehicle deformations on two crash tests on the same vehicle model. NHTSA crash test 3113 (left) and 3127 (right).

4.4 Determination of *A* and *B* stiffness coefficients from dynamic deformation

Residual deformation measurements can often be affected by errors due to the presence of bumpers or plastic parts. Especially in the new vehicle models, the bumpers are in plastic, and after the impact, they can return partially to their original shape. This phenomenon is very evident in low-speed impacts, where the vehicle may appear undeformed after the impact, even if the underlying structure is plastically deformed. In these cases it may be useful to derive the stiffness coefficients starting not from the residual deformations C but from the dynamic deformations X.

Dynamic deformations can be evaluated, for example, by arranging the acceleration curves of the vehicle during the crash, integrating twice to obtain the displacement (see Section 1.2) or, in the absence of these, also by analyzing the crash video, measuring moving a marker on the vehicle.

The use of dynamic deformations avoids the problem of measures distorted by the bumpers. The calculation of the stiffness coefficients can then be carried out by the procedure next, which is valid for a 100% frontal impact against an immovable rigid barrier.

Eq. (3.11), considering the dynamic deformation X, can be written as

$$V = b_1 X \tag{4.39}$$

where V is the impact speed, X is the measured dynamic deformation, from which is immediate to derive the b_1 coefficient

$$b_1 = \frac{V}{X} \tag{4.40}$$

and then, from Eq. (3.15) the B stiffness coefficient:

$$B = \frac{m}{L} {b_1}^2 \tag{4.41}$$

Knowing B, coefficient A can be calculated from Eq. (4.29).

In the case of a real accident, it is taken into account that the dynamic deformation is less affected by the curved shape of the bumper, the underlying reinforcement bar being practically straight in the vehicles. This makes it possible to estimate the dynamic deformation starting from the deformation of the crossbar, adding $\delta = A/B = b_0/b_1$.

4.5 Energy equivalent speed

Let us consider a vehicle of mass $m = 1000$ kg, initially stationary, which is hit by a nondeformable mobile barrier, of equal mass, at the speed V of 10 m/s. Let us assume that the impact is entirely inelastic, that is, the two vehicles reach an equal speed at the end of the impact. Since the barrier is nondeformable, all the kinetic energy is dissipated by the vehicle. From the momentum balance the common final speed can be derived as

$$\overline{V} = \frac{m}{m+m}V = \frac{1}{2}V = 5\ m/s$$

The energy balance is

$$\frac{1}{2}mV^2 = \frac{1}{2}m\overline{V}^2 + E_d$$

Substituting the determined final speed value, we get

$$E_d = \frac{1}{4}mV^2$$

This deformation energy can be written as

$$E_d = \frac{1}{2}m\left(\frac{V}{\sqrt{2}}\right)^2 = \frac{1}{2}m(7.07)^2$$

In other words the deformation energy of the vehicle is equal to the kinetic energy of the vehicle traveling at a speed of 7.07 m/s. This speed is said "EES," that is, is the speed equivalent, from the energetic point of view to the resulting deformation on the vehicle (Zeidler at al., 1985; Schreier and Nelson, 1987). The definition of EES is, therefore,

$$\text{EES} = \sqrt{\frac{2}{m}E_d} \tag{4.42}$$

In a collision between two vehicles, each of them is associated with an EES value, corresponding to the deformations found on the vehicle. For each deformation, for example, due to an intrusion of a tree on the side of the vehicle, an EES value can be evaluated.

In a plastic impact against a rigid and immovable barrier, all the initial kinetic energy is dissipated into deformation energy; EES is therefore often referred to as EBS, equivalent barrier speed, although the term EES emphasizes the link with the energy of speed, regardless of the existence of the barrier, and a crash test.

In crash tests made with a rigid but mobile barrier, it must be considered that part of the initial kinetic energy is transferred to the barrier; if M_b is the mass of the barrier and M_v is the one of the vehicles, from the conservation of the momentum and energy we get

$$\mathrm{EES} = \mathrm{ETS}\sqrt{\frac{M_b}{M_b + M_v}} \tag{4.43}$$

where ETS (equivalent test speed) indicates the vehicle's initial speed against the barrier.

In the fixed barrier account tests and with no, or negligible, elastic restitution, the EES coincides with the initial ETS test speed and with the speed variation ΔV of the vehicle.

If, however, the elastic restitution is not negligible, the vehicle has a nonzero speed at the end of the impact and, from the energy balance, is

$$\mathrm{EES} = \sqrt{\mathrm{ETS}^2 - \overline{V}^2} = \mathrm{ETS}\sqrt{1 - \varepsilon^2} \tag{4.44}$$

The following example highlights the difference between ΔV and EES.

Example

Let us consider a vehicle that hits a nondeformable tree at a speed V of 100 km/h and, deviated from the initial trajectory in a negligible way, continues driving at a speed $\overline{V}$ of 90 km/h.

In this case, $|\Delta V| = 100 - 90 = 10$ km/h.

The deformation energy, from the energy balance, results

$$E_d = \frac{1}{2}mV^2 - \frac{1}{2}m\overline{V}^2$$

and from Eq. (4.42), we get

$$\mathrm{EES} = \sqrt{V^2 - \overline{V}^2} = 43.6\ \mathrm{km}/h$$

If we assume $V = 50$ km/h and $\overline{V} = 40\ \mathrm{km}/h$, we get

$$|\Delta V| = 50 - 40 = 10\ \mathrm{km}/h$$

and

$$\mathrm{EES} = \sqrt{V^2 - \overline{V}^2} = 30\ \mathrm{km}/h$$

Therefore we can see that the EES is generally different from the ΔV, coinciding only in particular cases, for example, in the case of plastic impact and of zero postimpact speed.

Evaluating the EES of a vehicle, a comparison can be made with documented crash tests, on the same vehicle model, with damage located in the same area and with the same extension, knowing the amount of deformation energy. The condition that the two vehicles compared are of the same type, guarantees that they have the same structural behavior and therefore the same energy loss corresponds to the same level of deformations. If documented data of the same vehicle model are not available, at the expense of accuracy, it is possible to make the comparison using data from a vehicle that has a similar structural behavior, for example, identifying it within the same class in which they are grouped for coefficients A and B, or based on their wheelbase.

The comparison is of a qualitative type, based on the visual comparison of the damages, which must have a substantially similar extension, width and depth, or by comparing the deformation depth measurements. To obtain a reliable data the material used for the comparison should include an exhaustive photographic documentation and should be accompanied by measurements of the maximum deformation, the width of the deformed area, the degree of overlap, if there are intrusions in the passenger compartment, if and how much the engine has moved, the mass of the vehicle subjected to crashes, etc.

Fig. 4.9 shows an example of documented deformations, taken from a crash test with a 40% overlap on a fixed, rigid barrier, in which the deformation energy was 168.75 kJ, corresponding, given the mass of the vehicle of 1500 kg, to an EES of 54 km/h. It can see the magnitude of the direct and induced deformation, which also includes the extroversion of the right front side panel.

Knowledge of the mass of the crash-tested vehicle is necessary to calculate the deformation energy.

In case the vehicle has a different mass M from the one of available crash test, M_t (e.g., because there was a different number of persons on board), the EES must be corrected according to

$$\text{EES} = \text{ETS}_{\text{test}}\sqrt{\frac{M_t}{M}} \tag{4.45}$$

In case of impact between two vehicles, from Eq. (4.42), we get

FIGURE 4.9 Crash test on a fixed rigid barrier with 40% overlap and with a documented EES of 54 km/h. *EES*, Energy equivalent speed.

$$E_{d1} = \frac{1}{2}M_1\text{EES}_1^2$$
$$E_{d2} = \frac{1}{2}M_2\text{EES}_2^2 \quad (4.46)$$

Once the EES values determined for both vehicles, the total deformation energy is

$$E_d = E_{d1} + E_{d2} = \frac{1}{2}M_1\text{EES}_1^2 + \frac{1}{2}M_2\text{EES}_2^2 \quad (4.47)$$

In a collision between vehicles at high relative speed, in which the elastic restitution can be neglected, considering in first approximation a linear trend of the force with the deformation and considering that the forces exchanged between the vehicles must have the same values to one another, the energies of deformation of individual vehicles can be written as

$$E_{d1} = \frac{1}{2}FX_1$$
$$E_{d2} = \frac{1}{2}FX_2 \quad (4.48)$$

From Eqs. (4.48) and (4.46), we get

$$\frac{\text{EES}_1}{\text{EES}_2} = \sqrt{\frac{X_1M_2}{X_2M_1}} \quad (4.49)$$

The maximum deformations X, often indicated also with ETD, equivalent test deformation, represent the maximum deformations of the area of direct damage, measured along the direction of application of the resulting force in the impact.

Considering, as a first approximation, the contact forces F as linear functions of the deformation x, with a proportionality constant k, $F = kx$, and remembering Eqs. (3.5) and (3.12), Eq. (4.49) becomes

$$\frac{\text{EES}_1}{\text{EES}_2} = \sqrt{\frac{k_2M_2}{k_2M_1}} = \frac{M_2b_{12}}{M_1b_{11}} \quad (4.50)$$

From Eqs. (4.46) and (4.47), we can get

$$\text{EES}_1 = \sqrt{\frac{2E_d}{m_1(1 + (X_2/X_1))}}$$
$$\text{EES}_2 = \sqrt{\frac{2E_d}{m_2(1 + (X_1/X_2))}} \quad (4.51)$$

which allow deriving the EES values starting from the deformation energy, knowing the maximum dynamic deformations of the vehicles.

Eq. (4.50) allows, given the value of a vehicle's EES, to determine the value of the other, of which, for example, there are no comparison crash tests or data on structural characteristics available, such as in the case of impacts in areas where usually crash tests are not carried out.

In the case of frontal impacts between two vehicles, the expression (4.50) can be further simplified, since it can be considered that the stiffness of the front part of a motor vehicle is approximately proportional to its mass. This is because the vehicles must comply with specific structural requirements, for the safety of the occupants, which consist of a maximum value of acceleration and deformation consequent to a frontal impact. In a vehicle of lower mass, with the same acceleration, it must have a lower force, being $F = ma$; to obtain a lower force with comparable deformations the structure must have a lower stiffness, as shown in Fig. 4.10.

From this, it follows that in a collision between vehicles, since the forces exchanged are equal from the Newton's third law, the lightest vehicle undergoes the most significant deformations, as shown in Fig. 4.11.

Considering, therefore, that the stiffness of the front is approximately proportional to the mass of the vehicles, we have

$$\frac{X_1}{X_2} \cong \frac{M_2}{M_1} \tag{4.52}$$

Eq. (4.51), in case of frontal impacts, can be simplified as

$$\begin{aligned} \mathrm{EES}_1 &\cong \sqrt{\frac{2E_d}{m_1\left(1 + (m_1/m_2)\right)}} \\ \mathrm{EES}_2 &\cong \sqrt{\frac{2E_d}{m_2\left(1 + (m_2/m_1)\right)}} \end{aligned} \tag{4.53}$$

while from Eq. (4.49) we get

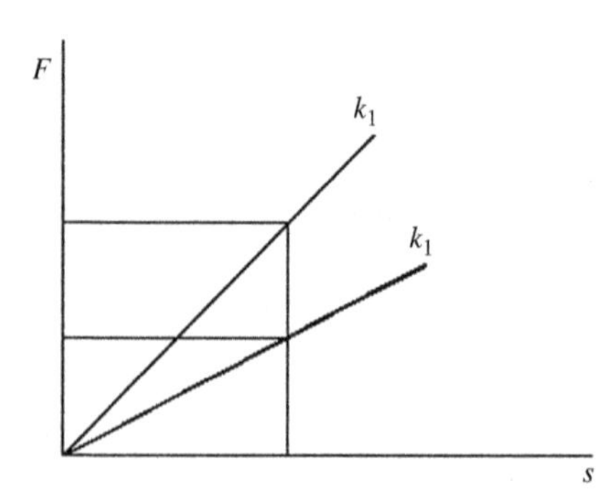

FIGURE 4.10 Force–deformation diagram for two vehicles with different stiffness k; with the same maximum deformation, to have a lower force, it is necessary to have a lower stiffness.

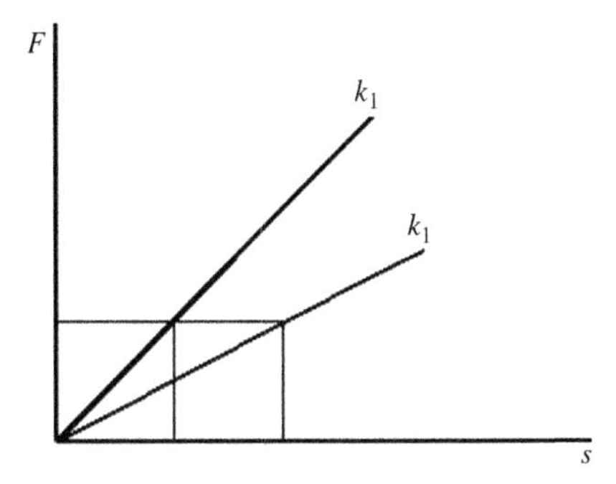

FIGURE 4.11 Force–deformation diagrams for two different stiffness k vehicles; with the same contact force, the vehicle with lower stiffness deforms more.

$$\frac{\mathrm{EES}_1}{\mathrm{EES}_2} \cong \frac{M_2}{M_1} \cong \frac{E_{d1}}{E_{d2}} \tag{4.54}$$

remembering Eq. (4.50), this means to have similar coefficients b_1 values for the two vehicles.

Since it must be

$$\int_0^t F dt = M_1 \Delta V_1 = M_2 \Delta V_2 \tag{4.55}$$

in the case of an impact against the barrier[2], for each vehicle, we can write

$$\mathrm{EES} = \frac{\left(\int_0^t F dt\right)_{\text{barrier } a}}{M} \tag{4.56}$$

and, similarly, in a collision between two vehicles:

$$\Delta V = \frac{\left(\int_0^t F dt\right)_{\text{collision } e}}{M} \tag{4.57}$$

Since in the case of a collision between two vehicles and in an impact against the barrier, to obtain the same deformation on a given vehicle, the contact forces must be equal, it must be

$$\frac{T_b}{T_c} = \frac{\left(\int_0^t F dt\right)_{\text{barrier } a}}{\left(\int_0^t F dt\right)_{\text{collision } e}} \tag{4.58}$$

and then dividing each member of Eqs. (4.56) and (4.57) and rearranging the terms, we get

$$\Delta V = \mathrm{EES} \frac{T_c}{T_b} \tag{4.59}$$

---

2. We always consider a substantially plastic collision, in which the restitution can be neglected, and then the speed change during the impact against the barrier is equal to the EES.

that shows the relationship between the speed variation in a collision between two vehicles and the *EES*. One can, therefore, say that in a centered collision between two vehicles, with the same mass and stiffness, the speed variation is equal to the *EES* for each vehicle. If the struck vehicle has a lower mass, the duration of the impact is less, and the speed variation of the colliding vehicle is lower than its *EES*, even if the deformations in the collision between the two vehicles and against the barrier have the same extension.

4.6 Triangle method

The triangle method combines the simplicity of the visual comparison, typical of the EES-based method, with the flexibility of the C_i-based method, which allows to evaluate the deformation energy on any deformation profile of the vehicle (Vangi, 2009a; Vangi and Begani, 2010, 2012, 2013).

This method overcomes some limits that are often encountered in the application of the abovementioned methods. In fact, in the case of C_i measurement, crash tests are often not available against a rigid barrier with a 100% offset to obtain stiffness coefficients A and B and those taken from the tables (by a class of vehicles) could be too general. In the case of EES, instead, the photographic documentation with similar deformations and on the same vehicle my not be available to make a comparison.

The triangle method is completely general and can be applied starting from photographic documentation with EES documented, on the same type of vehicles, relative to any deformation, even different from the one under study, obtained in any crash test or even in a real accident.

The method is based on the linearization of the damaged surface and approximation of any deformed part of the vehicle with a triangular, rectangular, or trapezoidal shape, as shown, for example, in Fig. 4.12.

For such damage configurations, it is possible to predetermine the analytical expression of the deformation energy as a function of only two parameters that characterize the shape of the damage: depth C and width L_d.

These parameters can be quantitatively estimated satisfactorily even from a visual analysis on suitable photographic documentation of the damage. For a correct evaluation of the damage depth C, it must be taken into account, even if in a qualitative way, the compared damage must be equivalent in terms of energy (the same damaged area at greater depth absorbs a more considerable energy, being the energy proportional to the square of the depth).

The procedure includes the following steps:

1. Evaluate the shape parameters of the damage on the reference vehicle, with known EES, to obtain the vehicle stiffness coefficients.

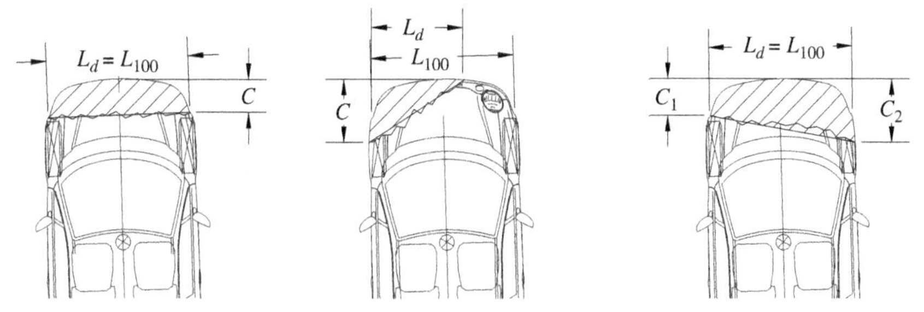

FIGURE 4.12 Approximation of damage on vehicles with areas of rectangular, triangular, or trapezoidal shape.

2. Evaluate the shape parameters of the damage on the vehicle in question, to obtain the known energy stiffness.

The mass–spring schematization, as proposed by Campbell (1974), is assumed based on empirical observation on the global behavior of vehicles during impacts against rigid barriers.

The energy E_a absorbed during the stage of compression Eq. (4.3), write in terms of b_0 and b_1 Campbell's coefficients, from Eq. (3.15) can be written as

$$E_a = \frac{m}{L}\sum_{1}^{n-1}\int_0^L \left(\frac{b_0^2}{2} + b_0 b_1 C_i + \frac{b_1^2 {C_i}^2}{2}\right) dl \tag{4.60}$$

For crashes with residual deformation more significant than a few centimeters, the elastic restitution is negligible, and the absorbed energy E_a during the compression stage is very close to the global energy dissipated E_d. In the following, $E_a = E_d$ is assumed.

The energy of deformation can be put in relation with the *EES* (4.33). In the triangle method the analytical *EES* expression is obtained in the cases in which the deformation has a simple geometry (triangular, rectangular, or trapezoidal), as a function of only the damage geometrical parameters C (depth) and L_d (width), as a ratio to total vehicle width L_{100}.

Triangle

If a triangle can approximate the deformed area on the vehicle (see Fig. 4.12), the deformation energy can be expressed as a function of the ratio between the width of the deformed area L_d and the width of the front L_{100}, the maximum deformation, and the stiffness coefficients characteristic of the vehicle b_0 and b_1.

$$E_d = L_d \frac{M}{L_{100}}\left(\frac{b_0^2}{2} + \frac{b_0 b_1 C}{2} + \frac{b_1^2 C^2}{6}\right) f \tag{4.61}$$

where f indicates the corrective factor to take into account that the PDOFs are not in general normal to the vehicle axis. Using, for example, Eq. (4.11), Eq. (4.61) becomes

$$E_d = \frac{M}{\cos^2(\mathrm{PDOF})}\frac{L_d}{L_{100}}\left(\frac{b_0^2}{2} + \frac{b_0 b_1 C}{2} + \frac{b_1^2 C^2}{6}\right) \tag{4.62}$$

By equating Eq. (4.33) to Eq. (4.62) a relation linking the product between the EES and the ratio between the width of the vehicle and the width of the deformed area is obtained, according to the maximum deformation detectable on the vehicle:

$$\mathrm{EES}\cos(\mathrm{PDOF})\sqrt{\frac{L_{100}}{L_d}} = \sqrt{b_0^2 + b_0 b_1 C + \frac{b_1^2 C^2}{3}} \tag{4.63}$$

Rectangle

If a rectangle can approximate the deformed area on the vehicle (see Fig. 4.12), it is possible to express the deformation energy as a function of the deformation depth C of the rectangle, the width of the deformed area $L_d = L_{100}$ (constant along the entire deformed perimeter), and the stiffness coefficients characteristic of the vehicle b_0 and b_1:

$$E_d = \frac{M}{\cos^2(\mathrm{PDOF})}\left(\frac{b_0^2}{2} + b_0 b_1 C + \frac{b_1^2 C^2}{2}\right) \tag{4.64}$$

Recalling Eq. (4.33), we obtain

$$\mathrm{EES}\cos(\mathrm{PDOF}) = \sqrt{b_0^2 + \frac{b_0 b_1 C}{2} + b_1^2 C^2} = b_0 + b_1 C \tag{4.65}$$

Trapezium

If the deformed area on the vehicle can be approximated by means of a trapezoid (see Fig. 4.12), the deformation energy can be expressed as a function of the width of the deformed area $L_d = L_{100}$, the maximum deformations C_2 and minimum $C_1 = kC_2$ $(k \le 1)$, and of the stiffness coefficients characteristic of the vehicle b_0 and b_1:

$$E_d = \frac{M}{\cos^2(\mathrm{PDOF})}\frac{L_d}{L_{100}}\left(\frac{b_0^2}{2} + \frac{b_0 b_1 C(1+k)}{2} + \frac{b_1^2 C^2(1+k+k^2)}{6}\right) \tag{4.66}$$

To reduce the number of parameters in Eq. (4.66), it is possible to approximate the area of the trapezium with the area of a triangle, equivalent from the point of view of the dissipated energy, having the same width and as length C the sum of the major deformation C_2 and the minor deformation C_1, multiplied by a factor of 0.7 (see Fig. 4.13). This factor gives the two areas, trapezoidal and triangular, analogous as dissipated energy.

$$C = C_2 + 0.7C_1 \tag{4.67}$$

At this point, we can express the deformation energy through (4.62) and obtain an expression for the EES analogous to Eq. (4.65), where we have $L_d = L_{100}$:

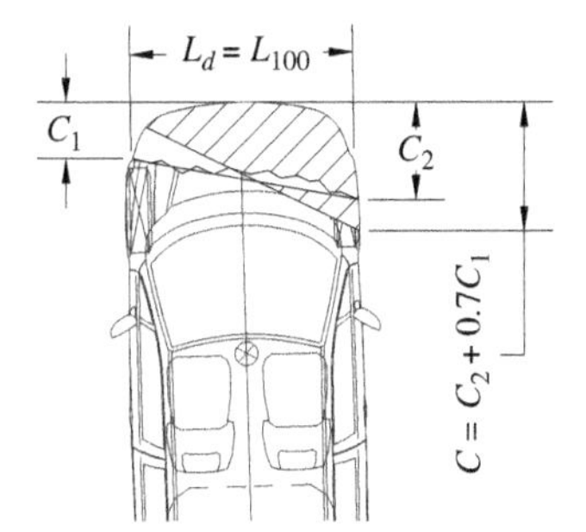

FIGURE 4.13 Approximation of the trapezoidal area with a triangular equivalent for the dissipated energy.

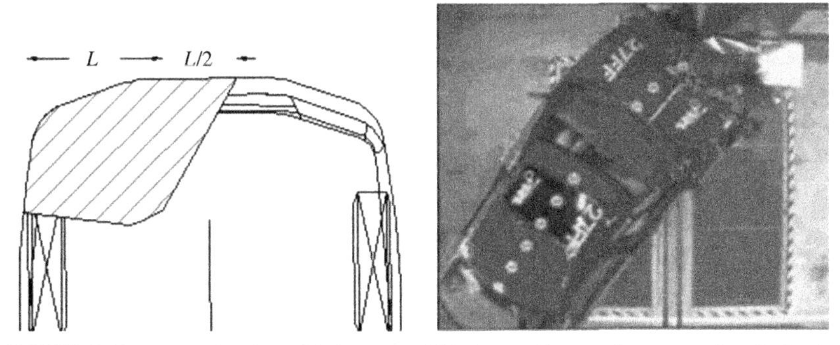

FIGURE 4.14 Approximation with the trapezoidal shape of impact damage against the barrier with 40% offset.

$$\text{EEScos(PDOF)} = \sqrt{b_0^2 + b_0 b_1 C + \frac{b_1^2 C^2}{3}} \tag{4.68}$$

Damage resulting from tests with 40% offset

The deformation resulting from a crash test against barrier or impactor with 40% offset is also analyzed, as it is a widespread test and therefore it can happen to have images of the damage with documented EES. The case of impact against a rigid or deformable (e.g., EuroNCAP tests) barrier is distinguished.

In the case of a crash test against a barrier with an offset of 40%, the deformation that occurs on the vehicle has a trapezoidal pattern, as shown in Fig. 4.14 (different from the trapezoidal shape examined earlier).

An area of direct damage characterizes this deformation, deriving from the contact between the front of the vehicle and the barrier, of a width equal to 40% of the width of the frontal $L = 0{,}4\ L_{100}$ and an area of indirect damage, whose width can be approximated, for almost all vehicles, equal to $L/2$.

The deformation energy of the vehicle (which is denoted by E_{dA}, to distinguish it from the globally dissipated E_d, which also includes that eventually dissipated by the deformable barrier) can then be expressed as the sum of rectangular damage plus a triangular one:

$$E_{dA} = M\frac{L}{L_{100}}\left(\frac{b_0^2}{2} + b_0 b_1 C + \frac{b_1^2 C^2}{2}\right) + M\frac{L/2}{L_{100}}\left(\frac{b_0^2}{2} + \frac{b_0 b_1 C}{2} + \frac{b_1^2 C^2}{6}\right) \tag{4.69}$$

The correction factor f due to the inclination of the forces with respect to the axis of the vehicle is neglected since typically the PDOF in this type of test is of a few degrees, and factor f can be considered unitary.

Since $L = 0.4\ L_{100}$, the previous relation can also be written as

$$E_{dA} = M\left(0.3b_0^2 + \frac{b_0 b_1 C}{2} + \frac{0.7 b_1^2 C^2}{3}\right) \tag{4.70}$$

Deriving the relationship that links the EES to the maximum deformation C on the vehicle, it is necessary to distinguish between the case of nondeformable and deformable barriers.

In the case of a nondeformable barrier the EES of the vehicle is equal (neglecting the elastic restitution) to the impact speed of the crash test. In the case of a deformable barrier part of the kinetic energy possessed by the vehicle dissipates in deformation of the barrier; the dissipated energy by the barrier is

$$E_{dB} = \frac{1}{2} k_B X_B^2 \tag{4.71}$$

with k_B barrier stiffness and X_B displacement (deformation) of the barrier. This displacement, in the real case of nonuniform barrier deformation, corresponds to a displacement calculated as an average weighted on the deformation depth of the barrier.

The deformation energy of the vehicle is

$$E_{dA} = \frac{1}{2} k_A X_A^2 \tag{4.72}$$

with obvious meaning of the symbols and the same considerations for the deformation.

Since at every moment of the impact the force exchanged is the same for the vehicle and the barrier, we have

$$k_B X_B = k_A X_A \tag{4.73}$$

From energy conservation, we have

$$\frac{1}{2} M V^2 - \frac{1}{2} M \overline{V}^2 = E_d \tag{4.74}$$

from which, recalling the definition of the restitution coefficient and considering that the barrier is fixed:

$$E_d = \frac{1}{2} M\ V^2 (1 - \varepsilon^2) \tag{4.75}$$

Then the energy loss E_{dA}, is

$$E_{dA} = \frac{1}{2}k_A X_A^2 = \frac{1}{2}\, M\, V^2\left(1 - \varepsilon^2\right) - \frac{1}{2}k_B X_B^2 \tag{4.76}$$

from which, by replacing k_B derived from Eq. (4.73) and with some passages, we get

$$E_d = \frac{1}{2}k_A X_A^2\left(1 + \frac{X_B}{X_A}\right) = \frac{1}{2}MV^2(1 - \varepsilon^2) \tag{4.77}$$

The dissipated energy by the vehicle is, by comparing Eq. (4.72) with Eq. (4.77)

$$E_{dA} = \frac{1}{2}MV^2\alpha^2 \tag{4.78}$$

with

$$\alpha^2 = (1 - \varepsilon^2)\left(\frac{X_A}{X_A + X_B}\right) \tag{4.79}$$

The EES of the vehicle, in an impact against a deformable barrier, is therefore equal to EES $= V\alpha$.

In the case of a crash test with a 40% offset against a deformable barrier, such as EuroNCAP, an average value of α equal to about 0.92 can be derived from the analysis of the ratio between X_A and X_B.

To obtain the relationship between EES and the maximum deformation C the following can be used instead of Eq. (4.33):

$$E_{da} = \frac{1}{2}MV^2\alpha^2 \tag{4.80}$$

where, in the case of an undeformable barrier, we put $\alpha = 0$, and we obtain

$$\text{EES} = V\alpha = \sqrt{\left(0.6b_0^2 + b_0 b_1 C + \frac{1.4b_1^2C^2}{3}\right)} \tag{4.81}$$

Eqs. (4.63), (4.65), and (4.81) yield a linear *EES* trend as a function of C, as shown in Fig. 4.15.

The three lines, besides a different slope, also show a slightly different intercept that can be put equal to b_0 as a first assumption. So all these above-mentioned equations can be approximated with a unique line with slope b_1, using *damage equivalent depth KC*:

$$\text{EES}\cdot\gamma = b_0 + b_1(KC) \tag{4.82}$$

where $\gamma = \sqrt{L_{100}/L_d}\cos(\text{PDOF})$ if the damage is triangular, $\gamma = \sqrt{\cos(\text{PDOF})}$ if the damage is rectangular, $\gamma = 1$ with 40% offset; b_0 and b_1 are Campbell's coefficients. The intercept b_0 represents the speed

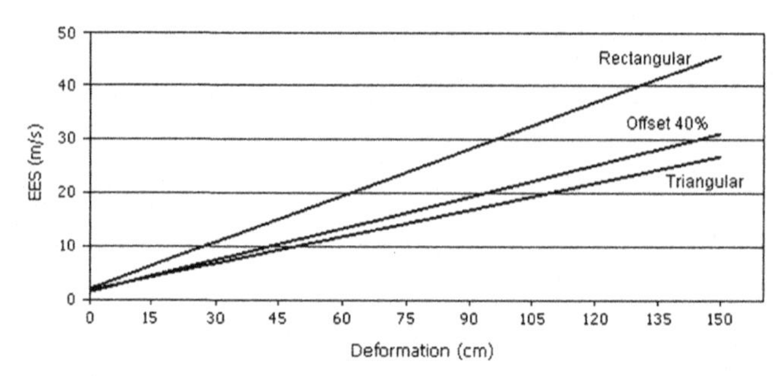

FIGURE 4.15 EES and damage depth C relation for various damage geometries. *EES*, Energy equivalent speed.

under which no permanent deformation is obtained in a crash against a rigid barrier, and as an indication, it can be put equal to 2 m/s. K is an appropriate shape parameter to allow for the difference in slope and intercept of the different lines. By comparison of Eqs. (4.65) and (4.82), it can be immediately derived that with rectangular damage the shape parameter K is equal to 1.

With triangular shaped damage or resulting from 40% offset crash, K parameter is calculated by way of minimizing, with the least square method, and as a function of b_1, the difference between EES values obtained from Eq. (4.82) and the one obtained from Eq. (4.63) or Eq. (4.81), therefore allowing for difference in slope and intercept of the lines. Nullifying the derivative of the sum S of the square deviations as a function of K:

$$\frac{\partial S}{\partial k} = \sum_i (\text{EES}_i \gamma_i - b_o - b_{1i} K C_i) = 0 \tag{4.83}$$

yielding the shape coefficient: $K = 1$ for rectangular damage, $K = 0.564$ for triangular damage, $K = 0.653$ for 40% offset crash.

In the general case in which the deformation can be assumed to have a trapezoidal geometry (see Fig. 4.12), EES can be expressed as (see Vangi, 2008)

$$\text{EES} = \sqrt{b_0^2 + b_0 b_1 (C_1 + C_2) + \frac{b_1^2 \left(C_1{}^2 + C_1 C_2 + C_2^2\right)}{3}} \tag{4.84}$$

that, as in the previous cases, can be approximated using Eq. (4.82), in which

$$C = C_2 - C_1 + \frac{C_1}{k} \tag{4.85}$$

where $K = 0.564$ as for the triangular damage shape.

The cases of damage with triangular or rectangular shape are special cases of the trapezoidal one, putting $C_1 = 0$ and $C_1 = C_2$, respectively.

Eq. (4.82) suggests that a deformation having a generic shape can be assimilated when calculating energy loss, to an equivalent deformation, with rectangular shape and depth equal to *KC*.

The use of the triangle method requires to find a value for the parameter b_1 of the vehicle under study using a similar vehicle with reference damage for which the *EES* value is known. According to the shape of the reference damage, Eq. (4.82) yields

$$b_1 = \frac{\mathrm{EES}\gamma - b_0}{KC} \tag{4.86}$$

Afterward, using Eq. (4.82), the desired EES for the vehicle under study can be calculated.

This can be done in a simplified way, using a unique expression:

$$\mathrm{EES}_O = \frac{1}{\gamma_O}\left[2 + \left(\frac{\mathrm{EES}_R\gamma_R - 2}{K_R C_R}\right) K_O C_O\right] \tag{4.87}$$

where subscript R indicates the reference damage, while subscript O indicates the damage under study. Eq. (4.87) represents the generalization of the triangle method and is applicable to any damage shape, with

$K = 1$ and $\gamma = \sqrt{\cos(\mathrm{PDOF})}$ for rectangular damage

$K = 0.564$ and $\gamma = \sqrt{(L_{100}/L_d)\cos(\mathrm{PDOF})}$ for triangular damage

$K = 0.653$ and $\gamma = 1$ with 40% offset.

Once the EES for the vehicle is known, from Eq. (4.33) the energy loss value can be obtained.

4.7 Triangle method with dynamic deformations

As for the determination of the coefficients A and B, also for the evaluation of the dissipated energy, it is possible to use dynamic deformations instead of residual ones.

For damage of a triangular shape, Eq. (4.3) can be written as

$$E_a = \int_0^{L_d} \frac{Bx^2}{2} dl \tag{4.88}$$

where x is the generic depth of the dynamic deformation, which can be expressed as a function of the maximum deformation X and the width of the deformed area L_d

$$x = \frac{X}{L_d} l \tag{4.89}$$

By integrating Eq. (4.88), we get

$$E_a = \frac{BX^2 L_d}{6} \tag{4.90}$$

and, from the definition of B (3.15):

$$E_a = \frac{m b_1^2 X^2 L_d}{6 L_{100}} \tag{4.91}$$

Neglecting, as done for the triangle method starting from residual deformations, the elastic energy recovered, we can equate Eq. (4.91) to Eq. (4.33) and obtain the EES:

$$\text{EES}\sqrt{\frac{L_{100}}{L_d}\cos(\text{PDOF})} = b_1 X \sqrt{\frac{1}{3}} \tag{4.92}$$

where the corrective factor has been added to take into account the PDOF.

For damage similar to a rectangle, Eq. (4.92), with similar considerations, becomes

$$\text{EES}\sqrt{\cos(\text{PDOF})} = b_1 X \tag{4.93}$$

while for damage with 40% offset, with deformable barrier:

$$\text{EES} = V\alpha = b_1 X \alpha \sqrt{\frac{7}{15}} \tag{4.94}$$

The values $\sqrt{1/3}$ and $\sqrt{7/15}$ obtained in Eqs. (4.92) and (4.94) are very close to the values $K = 0.564$ and $K = 0.653$ obtained in Section 4.6 for triangular damage and 40% offset, respectively. The difference between the values is due to the approximations made in (4.82), placing an equal intercept for all lines.

References

Brach, R., Brach, R., 2005. Vehicle Accident and Reconstruction Methods. SAE International, Warrendale, PA, 0-7680-0776-3.

Campbell, K.E., 1974. Energy Basis for Collision Severity. Society of Automotive Engineers, Inc., Warrendale, PA, SAE paper 740565.

CR81, 1981. Crash 3 User's Guide and Technical Manual. NHTSA, Washington, DC, Pub. No. DOT HS 805732.

Fonda, A.G., 1999. Principles of Crush Energy Determination. Society of Automotive Engineers, Inc., Warrendale, PA, SAE 1999-01-0106.

McHenry, B.G., 2001. The Algorithm of CRASH. McHenry software, Inc.

McHenry, R.R., McHenry, B.G., 1986. A Revised Damage Analysis Procedure for the CRASH Computer Program. McHenry Consultants, Inc., Cary, NC, SAE Paper 861894.

Neptune, J.A., Blair, G.Y., Flynn, J.E., 1992. A Method for Quantifying Vehicle Crush Stiffness Coefficients. Blair, Church & Flynn Consulting Engineers, SAE Paper 920607.

Nystrom, G.A., Kost, G., Werner, S.M., 1991. Stiffness Parameters for Vehicle Collision Analysis. Society of Automotive Engineers, Inc., Warrendale, PA, SAE Paper 910119.

Prasad, A.K., 1990. CRASH3 Damage Algorithm Reformulation for Front and Rear Collisions. TRC of Ohio Inc., East Liberty, OH, SAE Paper 900098.

Prasad, A.K., 1991. Energy Absorbed by Vehicle Structures in Side-Impacts. Transportation Research Center of Ohio, Inc., SAE Paper 910599.

Schreier, H.H., Nelson, W.D., 1987. Applicability of the EES-Accident Reconstruction Method With MacCAR©. Society of Automotive Engineers, Inc., Warrendale, PA, SAE870047.

Siddal, D.E., Day, T.D., 1996. Updating the Vehicle Class Categories. Engineering Dynamics Corp, SAE Paper 960897.

Smith, R.A., Noga, J.T., 1982. Accuracy and Sensitivity of CRASH. Society of Automotive Engineers, Inc., Warrendale, PA, SAE Paper 821169.

Tumbas, N.S., Smith, R.A., 1988. Measurement Protocol for Quantifying Vehicle Damage From an Energy Basis Point of View. Society of Automotive Engineers, Inc., Warrendale, PA, SAE Paper 880072.

Vangi, D., 2008. Accident Reconstruction. Firenze University Press, ISBN: 978-88-8453-783-6 (in Italian).

Vangi, D., 2009a. Simplified method for evaluating energy loss in vehicle collisions. Accid. Anal. Prev. 41 (3), 633–641.

Vangi, D., 2009b. Energy loss in vehicle to vehicle oblique impact. Int. J. Impact Eng. 36 (3), 512–521.

Vangi D., Begani F., 2010. The "triangle method" for evaluation of energy loss in vehicle collisions. In: IXX EVU Int. Conference, Prague.

Vangi, D., Begani, F., 2012. Performance of triangle method for evaluating energy loss in vehicle collisions. Proc. Inst. Mech. Eng., D: J. Automob. Eng. 226 (3), 338–347.

Vangi, D., Begani, F., 2013. Energy loss in vehicle collisions from permanent deformation: an extension of the "triangle method". Veh. Syst. Dyn. 51 (6), 857–876.

Woolley, R.L., Warner, C.Y., Tagg, M.D., 1985. Inaccuracies in the CRASH3 Program. Society of Automotive Engineers, Inc., Warrendale, PA, SAE Paper 850255.

Zeidler, F., Schreier, H.H., Stadelmann, R., 1985. Accident Research and Accident Reconstruction by the EES-Accident Reconstruction Method. Society of Automotive Engineers, Inc., Warrendale, PA, SAE Paper 850256.

Chapter 5

Crash analysis and reconstruction

Chapter Outline

Reconstructing an accident, particularly the impact phase between vehicles, requires an inverse engineering approach: from the effects of the impact, we have to going back to the causes, that is, from traces that remain of the accident we have to going back to the kinematic conditions before the impact. The result then leads to defining the maneuvers carried out by the participant and to understand the causes of the accident, generally attributable to the man, the road, or the vehicle and their interactions. Knowledge of the causes of the accident is of great importance both to understand accident rates to improve the design of vehicles and infrastructures and increase road safety and for legal purposes.

Like all reverse engineering problems, small uncertainties on the starting data can generate different incidental scenarios, that is, from the mathematical point of view, the system can be defined as ill-conditioned. This requires particular attention in the accident analysis, both in the single control of all initial data, taking into account their intrinsic uncertainty, both in the application of the models and in the assumptions and simplifications carried out.

Vehicle Collision Dynamics. DOI: https://doi.org/10.1016/B978-0-12-812750-6.00005-6

The previous chapters described the crash behavior of the vehicles and the models that can be used to analyze the impact phase between vehicle–vehicle and vehicle–barrier. These models, which can also be used for manual reconstruction, are the basis for the software for the reconstruction of traffic accident dynamics. In any case, the knowledge of these models, the physical laws and simplifying assumptions, must be well known to reconstruct the road accidents, both manually and with software. The correct interpretation and choice of the parameters of the reconstruction software require a thorough knowledge of what is at the base of their functioning.

The reconstruction of the impact phase and more generally of the incident as a whole can follow two different approaches: backward and forward analysis. In the first case the postimpact phase is analyzed first, obtaining the velocities and positions at the exit of the impact, then the impact phase, determining the speed and position of the vehicles at the moment of their impact, and finally the preimpact phase, with the analysis of any maneuvers carried out by the vehicles before reaching the impact. This approach is normally used in the manual reconstruction of the accident. In the second case the forward analysis is carried out assuming the initial conditions (positions and speeds) of the vehicles and simulating the successive instants, applying the motion equations and the impact models. The verification of the correctness of the initial data is carried out by verifying the correspondence of the simulation with the postimpact parameters. Generally, this analysis is carried out with software, which allows many simulations to be carried out, optimizing the input parameters to find a maximum correspondence with the available parameters.

The foremost computer software for simulations of automobile collisions (Simulation Model of Automobile Collisions—SMAC) was developed in 1974. As a preprocessor of the SMAC program, another program (Calspan Reconstruction of Accident Speeds on the Highway—CRASH) was also designed and was later developed as a stand-alone program CRASH3. In the recent years, based on the same algorithms, commercial software were designed to a great extent, such as EDSMAC, EDCRASH, m-smac, and later on in 1996 the programs WinSMAC and WinCRASH. In the mid-1990s, programs, such as CARAT3, PC-Crash, Virtual crash, Pro Impact, and others, were developed in Europe. Unlike the first the latter is based on impulsive models.

The fundamental supposition of the impulsive models is that the collision is instantaneous, that is, the vehicles do not change their positions during the collision phase. Such a supposition should be treated only as a first approximation. Furthermore, before using such models, a set of parameters should be known, such as the vehicle's position during the impact, the impact point, and the direction at which vehicles collided. Since in practice these data are often not available, some presumptions have to be adopted (e.g., impact duration and energy loss at impact). Otherwise, the determination of these parameters has to be left to the judgment of the user. In the models, based on vehicle deformations, contact forces between the vehicles are also considered in the dynamic equations. The contact forces are calculated based on the vehicle's mechanical properties. The main advantage of these programs is the possibility to allow a comparison between the calculated and actual deformation of the vehicles at collision.

A complete reconstruction of the impact phase between vehicles or between vehicles and obstacles generally also requires the reconstruction of the postimpact phase, that is, the phase in which the vehicles are no longer in contact with each other and continue their movement until the rest position. The input data for the reconstruction of these phases are generally constituted by the traces left by the motion of the vehicles on the ground or on the infrastructure (safety barriers, poles, etc.), the debris on the ground, the quiet positions of the vehicles, the position of the presumed point of impact, and finally the deformations on the vehicles.

From these data, the reconstruction process allows to generally obtain: the type and configuration of the collision, expressed by their relative position at the time of the first contact, the kinematic parameters of the collision (the initial velocities, the speed variations undergone by the vehicles following the impact, the accelerations), the severity of the impact.

For a successful usage of programs for collision simulations, reliable input data have to be provided. Among these, the photographs of the accident scene, photographs of the damage of the vehicles, and sketch of the accident scene are just the most important. Nowadays, excellent commercial software is available for processing photos of the accident scene. From accident photo-documentation the length of the skid marks can then be obtained, a digital model of the accident scene can be produced, and damage data of the vehicles can be collected. PhotoModeler, Pro Impact, and PC-Rect are examples of such programs.

5.1 Crash analysis

The models explained in the previous chapters can be schematically divided into impulsive models, based on the conservation of momentum and angular momentum, and models, based on the relationships between force and deformation of vehicles. Both types of models can be applied in backward or forward mode. In the forward analysis the impact models are the same as those used in the backward analysis, while the post- and preimpact phases are generally simulated through the equations of motion, assuming suitable models of pneumatic–terrain interaction for the forces exchanged with the ground. The present volume deals only with the impact phase, and therefore the simulation of the pre- and postimpact phases is not studied in depth, referring the reader to dedicated texts (Brach and Brach, 2005; Vangi, 2008).

The backward application of the two types of models requires the following steps:

- Impulsive models:
  - Determination of the impact configuration
  - Identification of the center of impact and contact plan
  - Analysis of the postimpact phase, to determine the speeds at the end of the impact
  - Calculation of preimpact speeds
  - Verification of results
- Models based on deformations:
  - Determination of the impact configuration
  - Evaluation of the direction of the resultant force [principal direction of forces (PDOF)]
  - Analysis of the postimpact phase, to determine the speeds at the end of the impact phase
  - Calculation of the deformation energy
  - Calculation of preimpact speeds
  - Verification of results

The various steps are summarized in the following subsections.

5.1.1 Impact configuration

Impact configuration means the relative position of the vehicles at the time of maximum crushing. In the event of a full impact, this corresponds to the end of the compression phase and the beginning of the return phase. In full impacts, during the compression phase, the vehicles are crushed, deforming one another. In this phase the rotations around the vertical axis and the direction of the original speed remain substantially the same. Generally, from the beginning of contact to the end of compression, the vehicles rotate a few degrees, as shown by way of example in Fig. 5.1.

With the linear models seen in Chapter 3, Models for the structural vehicle behavior, at the time of maximum compression corresponds the

FIGURE 5.1 Vehicles in frontal impact: on the left, the position at the first contact, on the right, at the maximum compression point. The yaw angle of the two vehicles during compression is negligible.

maximum forces exchanged. Using impulsive models to describe the impact phase, where the time is no present as a variable, this moment separates the impact from the postimpact. Vehicle speeds, that is, by the impulse received, change instantaneously from the values preceding the impact to the next ones.

For a correct application of the impulsive models and the reconstruction of the post- and preimpact phases, therefore, during the impact, the vehicles must be positioned in the configuration assumed at the end of the compression. For slide collisions an intermediate position, between the beginning of the contact and the end, can be assumed.

To position the vehicles in their relative impact configuration, it is useful to distinguish between direct and indirect deformations. The first are those caused by the application of forces, that is, those where the vehicle structures come into contact with one another. The latter are the deformations that are obtained without there being direct contact between the vehicles, as a consequence of direct deformations. To position the vehicles in the impact configuration, it is, therefore, necessary to match, as far as possible, the respective zones of direct deformation. Since the elastic restitution is generally negligible, the juxtaposition of the deformed vehicles is practically coincident with the position assumed by the same at the time of maximum compression.

Fig. 5.2 shows the juxtaposition of the 3D models of the vehicles, shown in Fig. 5.3, carried out with the Pro Impact software (http://www.atenaingegneria.it), in which the zones of direct deformation have been brought into contact.

5.1.2 Center of impact and contact plan

The center of impact is the point where the resultant of the contact forces can be considered applied. This point generally varies during the impact, as

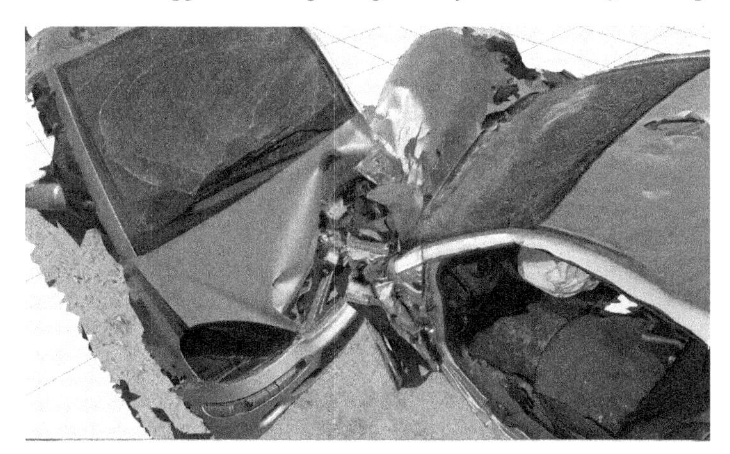

FIGURE 5.2 Juxtaposition of 3D models of vehicles in their position of maximum compression.

FIGURE 5.3 Example of 3D models of vehicles obtained with the Pro Impact software (http://www.atenaingegneria.it) via photogrammetry.

indicated in Section 2.1, and since impulsive models do not consider the time variable, the impact center must be identified, taking into account this variation. Considering the vehicles deformed and juxtaposed in their impact configuration, the center of impact can be approximately identified on the deformed surface of the vehicles, near of the most rigid structural elements. In fact, on these elements, such as the longitudinal members and the parts of the frame, the most significant contact forces are produced. This procedure requires a careful analysis of the deformations of the vehicles.

To identify the impact center and the impact plan in a more automatic way, it is possible, in the first approximation, to consider the outlines of the vehicles at the time of the first contact, as indicated in Fig. 5.4. From this position, since the speed variation is linear, with good approximation, one can consider the progress of each vehicle along its initial direction of motion, for a time equal to half the duration of the compression phase. The impact center can then be identified in the centroid of the overlapping area of the two vehicle shapes, as shown in Fig. 5.4.

The duration of the compression phase can be assumed to be equal to that provided, for example, by the linear model.

The contact plan can be taken as the one passing through the intersection points of the two vehicle shapes.

This procedure can be further simplified, to make it more easily automatizable, through geometric rules, assuming for the vehicles a rectangular shape (Kolk et al., 2016).

For each vehicle, their corners can be defined in the local vehicle coordinate system, as shown in Fig. 5.5. The coordinates of the corners are then transformed into the global coordinate system.

Advancing the vehicles along with the initial directions of travel for half the compression time, an overlap of the two rectangular profiles is obtained, as shown in Fig. 5.6. Knowing the stiffnesses of the parts of the vehicles coming into contact, it is possible, using the linear model Eq. (5.1), to evaluate the time of the compression phase.

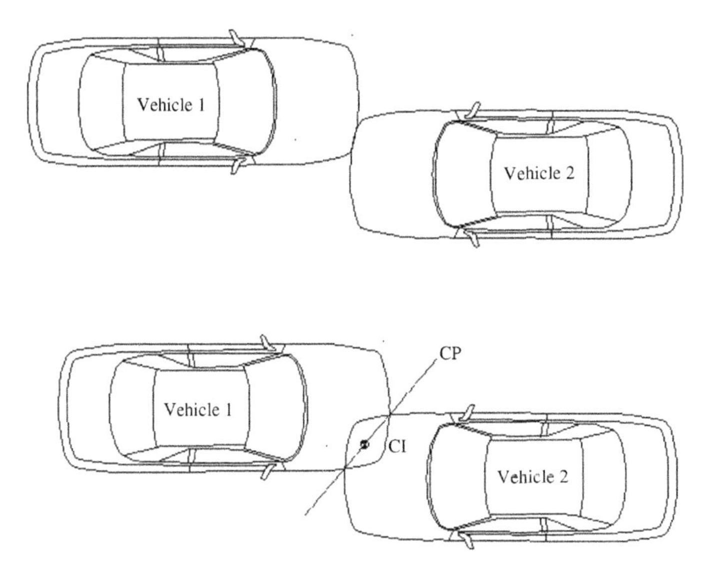

FIGURE 5.4 Vehicle shapes: at the time of first contact (top), at the time of maximum compression (bottom). The center and contact plan are also drawn.

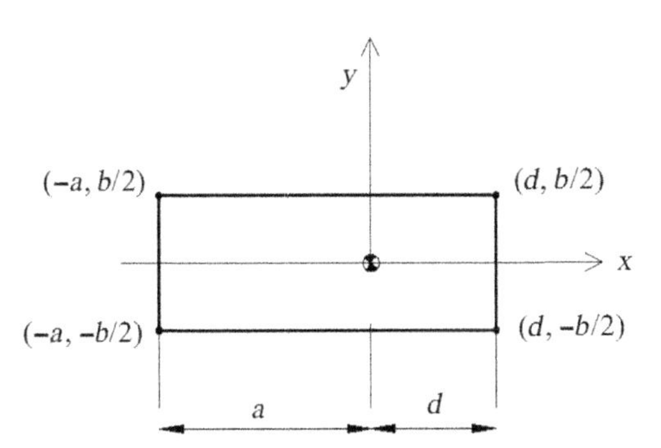

FIGURE 5.5 Local coordinates of the vehicle corners.

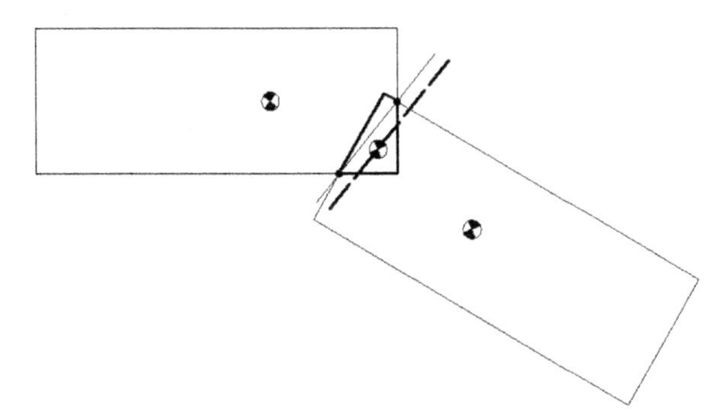

FIGURE 5.6 Overlap polygon between the vehicles and center and plane of impact as determined by geometrical rules.

$$t_x = \frac{\pi}{2}\sqrt{\frac{m_c^*}{k_c^*}} \tag{5.1}$$

The overlap is constituted by a polygon, and the corners are either corner of one of the colliding vehicles' exterior or the intersection of two edges of the vehicles' exterior. The polygon can be calculated based on the algorithm reported in Sutherland and Hodgman (1974). Knowing the coordinates of the n corners of the overlapping polygon, x_i and y_i, the coordinates of the center of impact, expressed as the centroid of the overlap, are given by

$$\begin{aligned} C_x &= \frac{1}{6A}\sum_{i=0}^{n-1}(x_i + x_{i+1})(x_1 y_{i+1} - x_{i+1} y_i) \\ C_y &= \frac{1}{6A}\sum_{i=0}^{n-1}(y_i + y_{i+1})(x_1 y_{i+1} - x_{i+1} y_i) \end{aligned} \tag{5.2}$$

where

$$A = \frac{1}{2}\sum_{i=0}^{n-1}(x_1 y_{i+1} - x_{i+1} y_i) \tag{5.3}$$

The contact plane can be defined interpolating a line through the intersection points between the rectangle vehicles' exterior and then translating in parallel such that it goes through the center of impact, as shown in Fig. 5.6.

5.1.3 Principal direction of forces

The identification of the PDOF can be made indirectly by observing the shape of vehicle deformations. In general, the PDOF can be approximated with the principal direction of the deformations (PDOD). Section 4.2.1 describes the procedure for identifying the PDOD, based on the superimposition of the deformed profile of the vehicle to the undeformed profile and joining the homologous points. The identification of the homologous points can be carried out by direct observation of the deformations, recognizing specific points in the deformed profile, for example, a headlight, a vehicle emblem, and a connection between two structures. Alternatively, based solely on CAD models of vehicles, homologous points can be identified by joining points at equal distance, measured on the deformed and undeformed profiles starting from an undeformed point, as shown in Fig. 5.7.

To determine the PDOF in a predictive manner (forward analysis), the results reported in Han (2015) can be referred to, assuming an impact plane as tangential direction t and a normal direction n, Ishikawa (1985, 1993, 1994), as shown in Fig. 5.8, or described in Kolk et al. (2016).

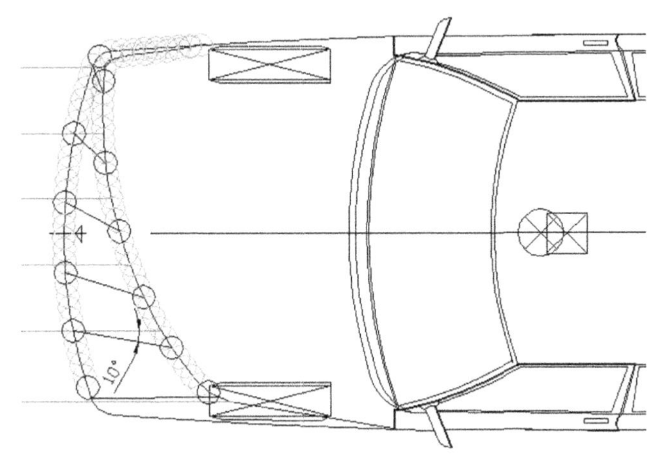

FIGURE 5.7 Identification of homologous points to determining the PDOD. *PDOD*, Principal direction of the deformations.

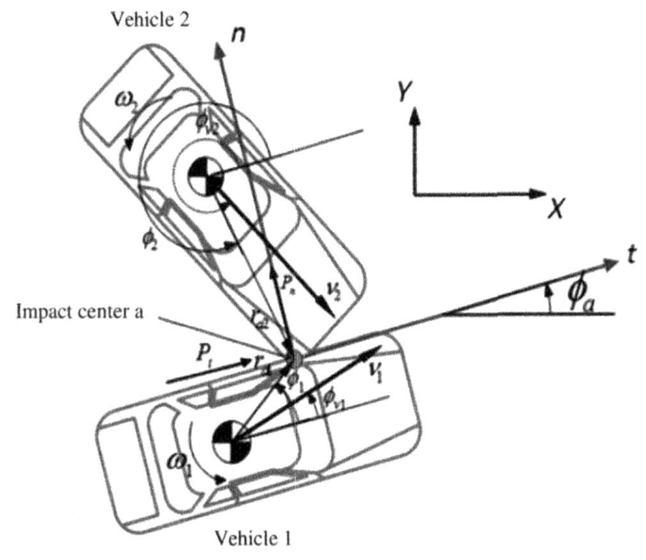

FIGURE 5.8 Vehicle planar impact: normal, n, and tangential, t, directions.

Referring to these directions, the following definitions can be given as follows:

- the coefficient of friction μ, expressed as the ratio between the component of impulse, tangential, and normal, during the impact Eq. (2.9)
- the speed ratio S_r, expressed as the ratio between the relative deformation speed V_{Rt}, along the tangential direction, and the relative slipping speed, along the normal direction, V_{Rn}

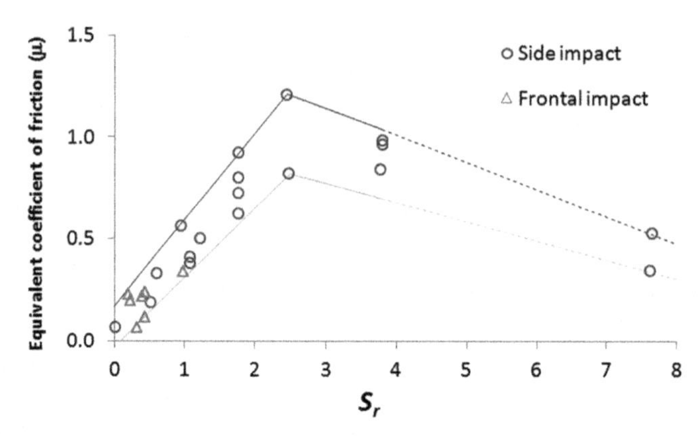

FIGURE 5.9 Equivalent coefficient of friction at impact surface.

$$S_r = \frac{V_{Rt}}{V_{Rn}} \tag{5.4}$$

Fig. 5.9 shows an empirical relationship between μ and S_r. Knowing the relative speed of impact between the two vehicles, and calculating the ratio between its tangential and normal components, μ can be obtained and then, the desired value of PDOF.

The scatter of experimental data results showed uncertainties on the PDOF in the range of ± 15 degrees. Other analyses of literature (Smith and Noga, 1982) confirm that the PDOF, for each vehicle, can vary by ± 20 degrees in relation to the subjective assessment of the impact plane.

5.1.4 Postimpact phase analysis

The kinematic analysis of the motion after the collision of vehicles is the first phase of the road accident backward reconstruction, after having defined the relative position of impact of the vehicles. To assess the distances traveled, and the rotations performed by vehicles during the postimpact phase, the point of impact on the road must be identified. Generally, the position of the point of impact is identified by the traces on the ground, such as skidding marks, debris, incisions, traces of liquids, or gum, and also from the information detected on the scene, such as the initial direction of travel and the maneuvers implemented.

In the postimpact phase the most significant motions are translations and yaws. These motions can occur with wheels free to roll or wheels blocked or partially blocked. Even if the driver is not braking before and after the impact, the deformations of the structures often block all or part of the rotation of some wheel. Generally, the starting data for the reconstruction of the accident are few and often uncertain. This makes useless, or even misleading, to reconstruct postimpact motion using models that require many data, such as steering angle, longitudinal and transverse friction coefficients, the trajectory traveled by the vehicle, and the stiffness of the suspension.

In most cases, therefore, the transfer of load between the wheels due to pitching, rolling, and yawing can be neglected, and the vehicle can be considered as a rigid body with 3 degreesof freedom: the two translations on the plane and the rotation around the vertical axis. The postimpact roto-translation movement of the vehicle can be described by integrating the motion equations with a small time step, considering the forces acting on the vehicle itself (Gillespie, 1992), starting from the distance and yaw angle traveled by the vehicle. For wheel-paving forces approximations, such as the circle or the ellipse of friction, can be used (Brach and Brach 2001; Pacejka, 2005). In this way, both free-wheeled and blocked or partially blocked motions can be described. The simplified model reported by Macmillan (1983), valid for locked wheels, allows uncoupling the equations for the calculation of the translation speed from those for the calculation of the angular speed, with an iterative procedure. The Vangi model (Vangi, 2013), which is also valid for blocked wheels, does not require iterations and produces comparable results with the direct integration of the motion equations.

The description of these models is outside the scope of the present text, which deals only with the vehicle impact phase.

5.1.5 Energy loss evaluation

The calculation of the deformation energy can be performed by applying one of the methods illustrated in Chapter 4, Energy loss. The classic method based on the deformation measurements, illustrated in Section 4.1, can be used, applying eventually the correction illustrated in Section 4.2, if the PDOF is different from zero. For this procedure, it is necessary to know the values of the stiffness coefficients A and B of the vehicle, which can be found from tables in literature (https://www.nhtsa.gov), determined by crash tests based on residual deformation measurements as shown in Section 4.3 or dynamic measurements, as shown in Section 4.4.

Alternatively, the deformation energy can be determined based on comparisons with deformed vehicles of which the energy equivalent speed (EES) is known, as illustrated in Section 4.5. In this case, vehicles identical to the vehicle under observation, or at least that they belong to the same stiffness class, and which visually have the same type and extent of deformations, must be taken as reference.

Finally, the deformation energy can also be determined using the Triangle method, described in Sections 4.6 and 4.7, both from residual and dynamic deformation measurements.

5.1.6 Preimpact velocity calculation

The calculation of preimpact velocities differs depending on whether methods based on momentum conservation or deformation-based methods are used.

In the first case the impulsive models shown in Section 2.5 are applied. In particular, the values of the normal and tangential pulse can be calculated as described in Section 2.5.1, assuming values for the restitution coefficient ε (see Sections 2.3 and 2.5.2) and for μ (see Section 2.7.2).

If, from an analysis of the motion before and after the impact, it is possible to assess the speed directions of the post and preimpact vehicles, then it is convenient to use the formulations reported in Section 2.5.3, which avoid having to estimate ε and μ.

If deformation-based methods are used to calculate the preimpact velocities, the modulus of the velocity variation vector ΔV of vehicles must be determined first. This can be done as described in Section 2.8. To determine the preimpact speeds the speed triangles for each vehicle must be considered. The pre- and postimpact velocity vectors and the vector ΔV must form a closed triangle.

If the module and the direction of the vector ΔV, which coincides with the PDOF, is known, the triangle closes making the vector sum of the postimpact speed with the ΔV, as shown in Fig. 5.10.

If, instead, the direction of the vector ΔV is unknown and we have only its module and the preimpact speed direction, the unknown quantities are determined by closing the speed triangle, as shown in Fig. 5.11.

In this case, two solutions are possible for the preimpact velocity vector, *OA* and *OB*, where points *A* and *B* are determined by the intersection of the circumference of radius ΔV with the straight line identifying the direction of the preimpact velocity. To determine which is the correct solution, it is also necessary to consider the triangle of the speeds of the other vehicle, remembering that the direction of the vector ΔV must be the same for the two vehicles, as shown in Fig. 5.12. In the figure, drawn the known parameters for both vehicles (postimpact speed vectors, preimpact directions, and circumferences of radius ΔV), the only correct direction for ΔV is the one that passes through points *A*.

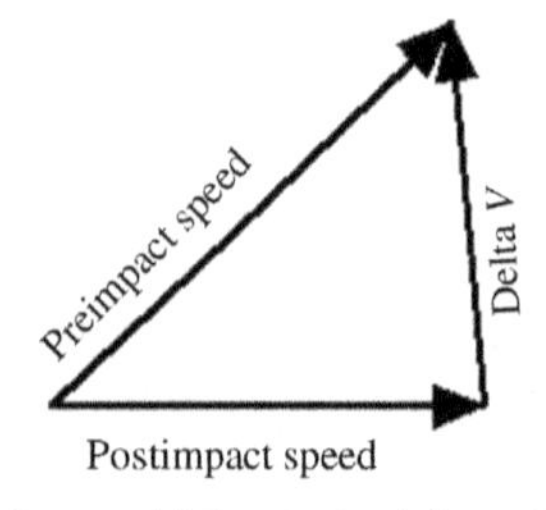

FIGURE 5.10 Speed triangle for a vehicle, obtained from the knowledge of the postimpact speed and the ΔV.

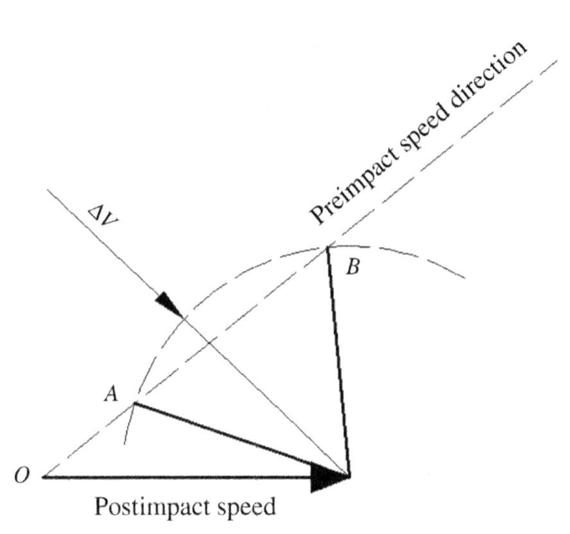

FIGURE 5.11 Determination of preimpact velocity, knowing postimpact velocity, ΔV module, and preimpact direction.

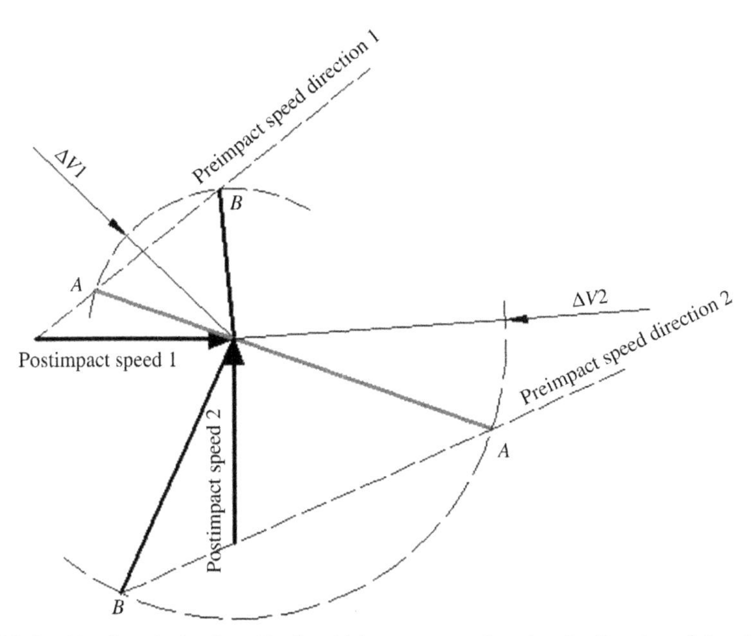

FIGURE 5.12 Speed triangles of both vehicles, constructed so that the direction of the ΔV is the same for both: only the intersection in A satisfies this criterion.

5.1.7 Results check

Often the starting data to reconstruct an impact are affected by errors, deriving from uncertainties on the data collected (traces, deformations, skid marks, etc.), on the parameters assumed and on the correct application of the models. Thus it is always necessary to carry out checks to make the kinematic values calculated congruent with all available data and usable models. The type of verification may vary depending on whether the calculation was performed with impulsive models or deformation-based models and depending on the available input data. However, the indispensable checks are those relating to the momentum conservation, the triangles of speed, and the energy balance.

Check 1—momentum conservation

The following scalar relationship can express the momentum conservation:

$$\frac{m_A}{m_B} = \frac{\Delta V_B}{\Delta V_A} \tag{5.5}$$

where m_A and m_B, and ΔV_A and ΔV_B are the masses and modules of velocity change vectors of vehicles *A* and *B*, respectively.

Eq. (5.5), if the forces exchanged between vehicles and road surface are negligible (isolated system), applies to all types of crashes regardless of velocities and impact configurations involved (full impacts, sliding impacts, etc.). This assumption is generally acceptable when impact forces are prevalent, that is, when the crash occurs at a sufficiently high closing speed $V_r > 10$ km/h.

The first check on calculated data is based on the application of Eq. (5.5), considering the real masses at the instance of the accident; hence, masses of passengers and loads must also be accounted for. If Eq. (5.5) is not satisfied, then data are not congruent, and the incongruity may depend on the masses, or it may be a consequence of using an incorrect accident reconstruction model or procedure.

Check 2—compatibility between pre- and postimpact velocities

The calculation of speed variation ΔV and closing speed V_R in the models based on vehicle deformations relates only to the modulus; thus other information regarding the vectors are not available. For a generic vehicle, velocity change, initial velocity, and final velocity vectors (ΔV, V, and $\overline{V}$, respectively) form a closed triangle, because the vector sum, $V_i + \Delta V_i = \overline{V}_i$, must apply. Thus a second check for the data coherence is to assess that the three vectors compose a triangle.

To apply the check the possible uncertainties in the vector direction and ΔV values should be considered. An example of the inconsistency in the velocity triangle is shown in Fig. 5.13, for the cases $|\Delta\alpha| < 90$ degrees and 90 degrees $\leq |\Delta\alpha| \leq 270$ degrees.

If the check is not satisfied, then data are not congruent; this may derive from an incorrect application of the physical model or incorrect data employed (e.g., in pre- or postimpact directions).

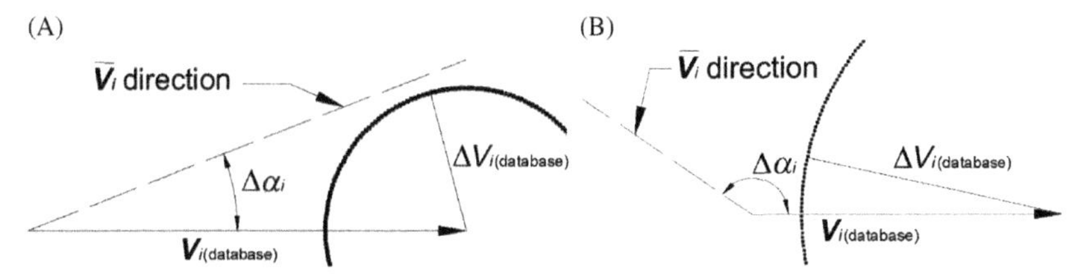

FIGURE 5.13 Conditions of velocity triangle inconsistency: (A) $|\Delta\alpha| < 90$ degrees; (B) 90 degrees $\leq |\Delta\alpha| \leq 270$ degrees.

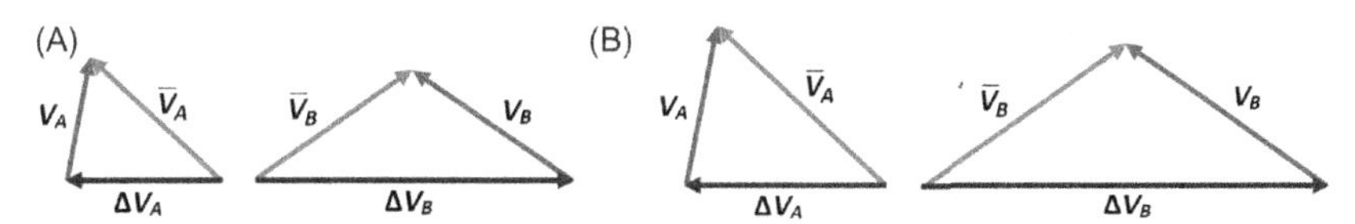

FIGURE 5.14 Velocity triangles satisfying checks 1 and 2, but the case (A) has a different scale factor from the case (B).

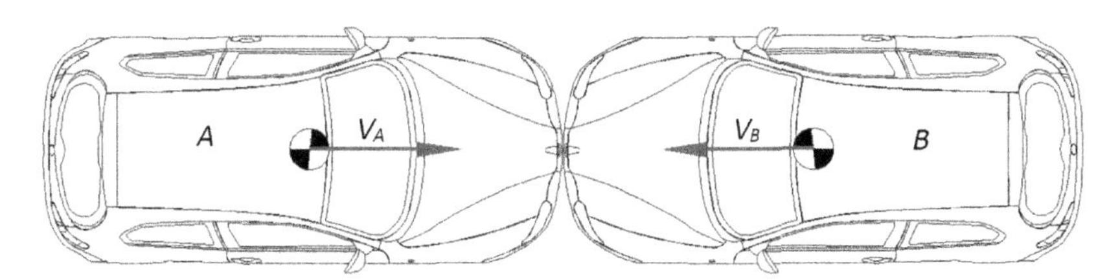

FIGURE 5.15 Collision between two identical vehicles in which only the extent of deformation can provide adequate indications of reconstruction correctness.

Check 3—energy conservation

Consider a case in which both checks 1 and 2 are satisfied, that is, the ratio between ΔV of vehicles A and B is congruent with their mass ratio, and the vectors form a closed triangle, as shown in Fig. 5.14A. For the same accident, Fig. 5.14B presents same kinematics parameters but scaled at a specific factor. Even in this case, checks 1 and 2 are satisfied, but the absolute impact velocity of the two vehicles is different. Thus it is necessary to impose an additional check on the coherence of kinematics data and those related to the accident (e.g., the consequences of the impact on vehicles).

Consider the example shown in Fig. 5.15, in which two identical vehicles (equal masses) collide at equal velocities. Considering the crash as perfectly plastic, the final velocities are null. The application of Eq. (5.5) yields the relationship $V_A = V_B = V$, which can be satisfied by infinite combinations of kinematic parameters determined as congruent through checks 1 and 2. Different initial velocities imply different permanent deformations. A check on the deformation energy, E_d, allows defining the correct values, congruent with the vehicle's deformations. For the energy conservation law (neglecting the eventual energy associated with vehicle rotation), the following relationship must hold

$$\Delta E_c = \frac{1}{2}\left(m_A V_A^2 + m_B V_B^2\right) - \frac{1}{2}\left(m_A \overline{V}_A^2 + m_B \overline{V}_B^2\right) = E_d \tag{5.6}$$

If the characteristics of occupants are unknown, an average mass of 70 kg for each occupant can be assumed to compute the actual mass of each vehicle.

To apply Eq. (5.6), it is necessary to know the pre- and postimpact velocities of the two vehicles, as well as the deformation energy. If it is assumed an error $t_e = 3$ km/h (Vangi, 2009) in estimating the EES by using approximate methods, then the following relationship can be deduced

$$\begin{aligned} E_{d\_\min} &= \frac{1}{2} m_A(\mathrm{EES}_A - t_e)^2 + \frac{1}{2} m_B(\mathrm{EES}_B - t_e)^2 \\ E_{d\_\max} &= \frac{1}{2} m_A(\mathrm{EES}_A + t_e)^2 + \frac{1}{2} m_B(\mathrm{EES}_B + t_e)^2 \end{aligned} \tag{5.7}$$

Therefore the check on kinetic energy can be expressed as follows:

$$E_{d\_\min} \leq \Delta E_c \leq E_{d\_\max} \tag{5.8}$$

If the check Eq. (5.8) is not satisfied, a coherence error between kinematic data and impact consequences exists.

5.2 Example

In this chapter, we analyze and reconstruct the impact phase between two vehicles, applying the various models and techniques seen in the previous chapters, and carrying out the relevant checks and verifications. The steps described in Section 5.1 will be followed.

A real accident between two vehicles is considered, a Mercedes and a Rover, coming into impact at a 90 degrees intersection. The characteristics of the vehicles are shown as follows:

- Range Rover Evoque:
  - Mass m: 1891 kg
  - Moment of inertia I: 3066 kg m^2
  - Vehicle length (m): 4.36
  - Vehicle width (m): 1.99
  - Wheelbase (m): 2.67
  - Overhang(m): 0.80
  - Center of gravity-front axle distance (m): 1.23
- Mercedes ML 350:
  - Mass m: 2274 kg

- Moment of inertia I: 4662 kg m^2
- Vehicle length (m): 4.78
- Vehicle width (m): 1.91
- Wheelbase (m): 2.92
- Overhang (m): 0.87
- Center of gravity-front axle distance (m): 1.20

5.2.1 Impact configuration and point of impact

The impact configuration can be deduced from the vehicles deformation analysis, represented in Figs. 5.16 and 5.17. The front of the Rover is globally deformed, with an accentuation of the "V" shaped introflexion in the center-left part, which finds a perfect match in the deformation of the left side of the Mercedes. In fact, on the Mercedes side, you can see that the B pillar, with high stiffness, compared to the doors, offered greater resistance and caused the "V"-shaped recess on the front of the Rover. The side of the Mercedes shows a greater introflexion on the rear door compared to the front one.

By virtually matching the deformed profiles of vehicles, an impact configuration is obtained as shown in Fig. 5.18.

FIGURE 5.16 Mercedes side deformations.

FIGURE 5.17 Rover front deformations.

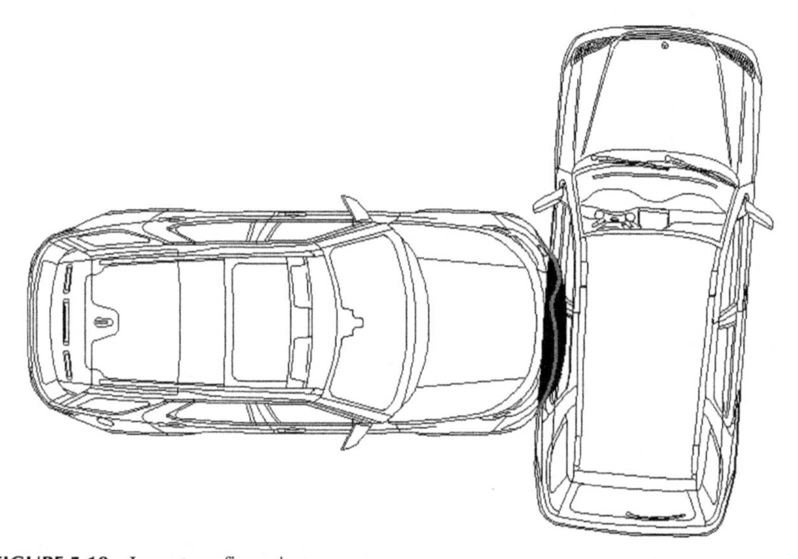

FIGURE 5.18 Impact configuration.

This relative position of the vehicles corresponds, with good approximation, to the configuration at the moment of maximum compression. Fig. 5.19 shows the moment of impact between the two vehicles, taken from a video obtained by a video surveillance camera. The vehicle impact configuration is entirely congruent with the relative position taken from the deformation analysis.

5.2.2 Center of impact an contact plane

The center of impact represents the point at which the resultant of the contact forces between the vehicles can be applied during the entire duration of the impact. Since the Mercedes B pillar represents the most rigid point of the

FIGURE 5.19 Position of vehicles at the time of their impact, the frame taken from a video surveillance video camera.

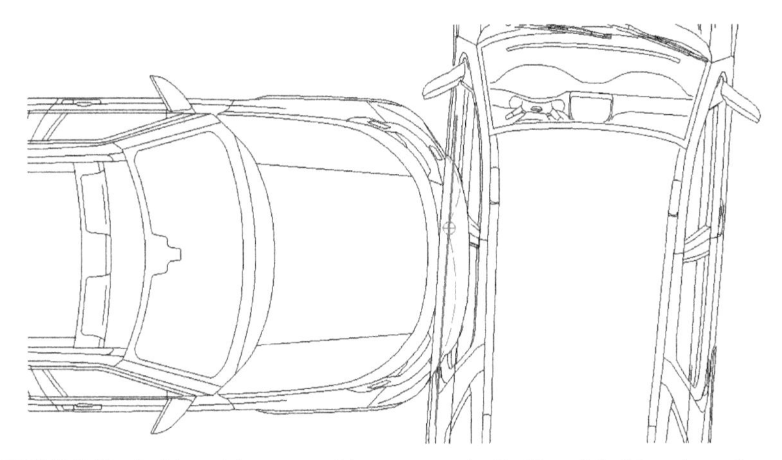

FIGURE 5.20 Position of the center of impact, near the B pillar of the Mercedes and contact plane, parallel to the Mercedes side.

side part involved in the impact, it is reasonable that at that point have developed the most significant contact forces, and this can be considered as a center of impact.

The contact plan can be assumed to coincide with the side surface of the Mercedes. Fig. 5.20 shows the position of the center of impact and contact plan, thus evaluated.

By applying the automatable process, described in Section 5.1.2, the center of impact would be located in the centroid of the overlapping area of the two vehicle shapes, as shown in Fig. 5.21. This position results in correspondence with the rear door, in an unlikely position.

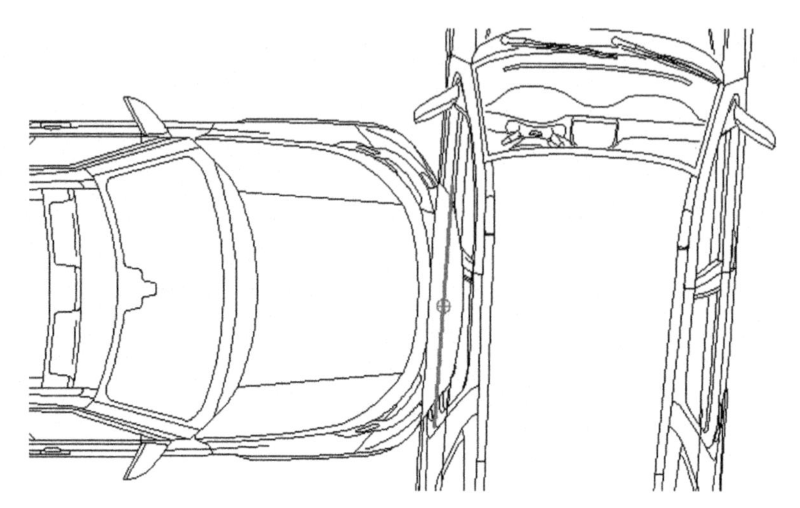

FIGURE 5.21 Position of the center of impact and contact plan according to the automatable procedure described in Section 5.1.2.

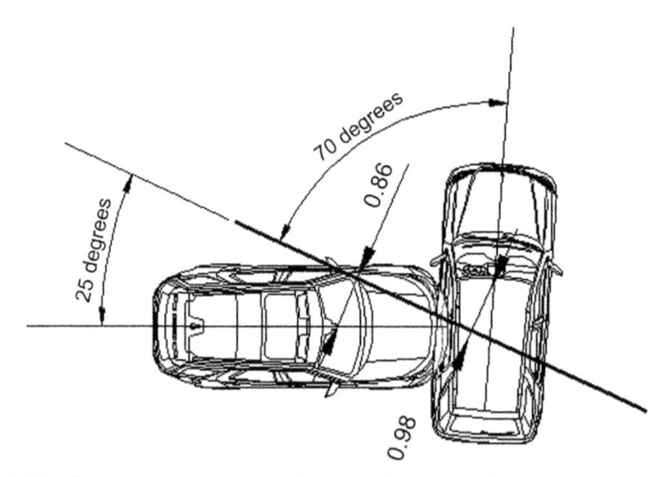

FIGURE 5.22 Estimate of the principal direction of the forces, based on the deformations on the vehicles.

5.2.3 Principal direction of forces evaluation

The direction of the resultant of the forces can be qualitatively assessed by the trend of the deformations of the vehicles. From Figs. 5.16 and 5.17, we see that the front of the Rover, in addition to the introflection, shows a twist toward the left of the car, and also the deformation of the Mercedes shows a shift of the rear door plates slightly toward the rear. The main direction of the forces can then be estimated as shown in Fig. 5.22, which corresponds to a PDOF of about 25 degrees for the Rover and about 20 degrees (90 − 70 degrees) for the Mercedes.

The arms of the resultant forces can be estimated to be 0.86 and 0.98 m for the Rover and Mercedes, respectively.

The PDOF can be verified later, starting from the calculated impact speeds and the conservation of the momentum.

5.2.4 Postimpact analysis

The position of the point of impact between the vehicles, with respect to the road, can be deduced in Fig. 5.19. Fig. 5.19 shows the positions of the vehicles at the moment of impact, the rest positions, and the vehicles centers of gravity trajectories of the postimpact.

From the analysis of the video of the accident, it is possible to evaluate the postimpact velocities of the vehicles and the initial and final angles of entry and exit with respect to a reference system $n - t$, chosen with the direction t aligned with the impact plane.

For Rover we have

- initial direction θ 185 degrees,
- final direction $\overline{\theta}$ 201 degrees,
- postimpact speed $\overline{V}$ 24 km/h, and
- final angular speed $\overline{\omega}$ 2 rad/s.

For the Mercedes, analogous, we have

- initial direction θ 270 degrees,
- final direction $\overline{\theta}$ 246 degrees,
- postimpact speed $\overline{V}$ 35 km/h, and
- final angular speed $\overline{\omega}$ 2.2 rad/s.

It is useful to point out that, although the values of the postimpact angular speeds are the same, the angles of rotation traveled by the vehicles are very different; the Mercedes has rotated 125 degrees while the Rover has only 22 degrees. This is because the rear of the Mercedes, as a result of the impact, has lifted off the ground and the rotation took place solely on the front wheels. Conversely, after a first bump, the Rover relied heavily on the four wheels, whose contact forces with the pavement stopped the rotation in progress. Furthermore, the trajectories of the vehicle's center of gravity are broken lines, as the Mercedes, after reaching the end of the rotation, moved backwards due to the slope of the road, while the Rover, after an initial shift, continued on a more deviated trajectory, aligned with the yaw angle reached.

5.2.5 Energy loss calculation

The deformation energy can be evaluated by measuring the extent of the deformations, or by evaluating the EES of the vehicles by comparison with known cases.

Measurement of the extent of deformations

First of all, it is necessary to determine the stiffness constants of vehicles. For Mercedes, the EuroNCAP crash test on a lateral pole, which the maximum deformation is shown in Fig. 5.23, can be considered a reference deformation. The crash test EES value is 29 km/h (8.1 m/s), which corresponds to a deformation energy value of 71.5 kJ.

On the CAD model of the vehicle, it is possible to redesign the reference deformation of the vehicle, as shown in Fig. 5.24, and apply the triangle method.

FIGURE 5.23 Side crash crash test for the Mercedes, taken as a reference to estimate the lateral stiffness of the vehicle.

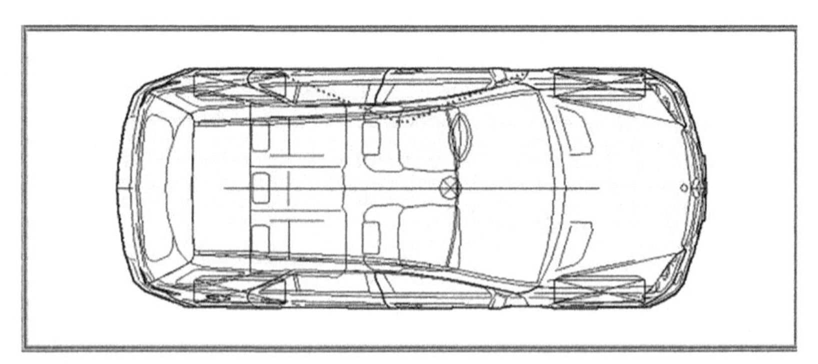

FIGURE 5.24 CAD image of the Mercedes vehicle, with the deformed profile drawn from the comparison with the crash test of Fig. 5.23.

The deformation can be assimilated to a triangle, whose maximum depth is equal to about 41 cm, the width of the deformed area is equal to about 42% of the length of the vehicle and PDOF = 0, from which we obtain

$$\gamma = \sqrt{\frac{L_{100}}{L_d}\cos(\text{PDOF})} = 1.54$$

and a b_1 value:

$$b_1 = \frac{\text{EES}\gamma - b_0}{kC} = \frac{8.1 \cdot 1.54 - 1}{0.564 \cdot 0.41} = 49.4$$

The deformation of the object vehicle is schematically shown in Fig. 5.25. The deformed profile of the side can be assimilated to a triangular deformation, whose maximum introflection is about 22 cm, the extension is equal to about 46% of the length of the vehicle and the PDOF = 20 degrees.

With these data, you get

$$\gamma = \sqrt{\frac{L_{100}}{L_d}\cos(\text{PDOF})} = 1.43$$

and, with the b_1 already get, we obtain

$$\text{EES}_O = \frac{1}{\gamma^O}[1 + b_1 k_O C_O] = \frac{1}{1.47}[1 + 49.4 \cdot 0.564 \cdot 0.22] = 4.38\,\text{m/s} = 15.8\,\text{km}/h$$

corresponding to a Mercedes energy loss of 21.8 kJ.

For the Range Rover, similarly to Mercedes, we can consider the EuroNCAP crash test at 40% offset against the deformable barrier, shown in Fig. 5.26. The EES value of the crash test is 59 km/h, which corresponds to an energy loss of 242.5 kJ.

On the CAD model of the vehicle, it is possible to redesign the deformation of the vehicle, as shown in Fig. 5.27 to assess the extent of the intrusion.

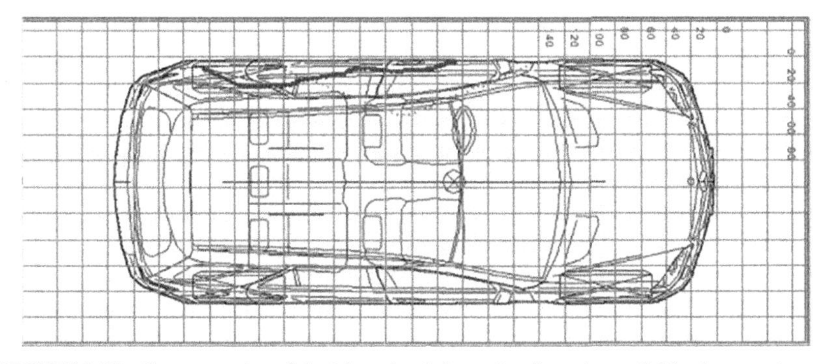

FIGURE 5.25 Representation of the Mercedes deformation from the available photographs.

FIGURE 5.26 Range Rover crash test taken as a reference to estimate the frontal stiffness of the vehicle.

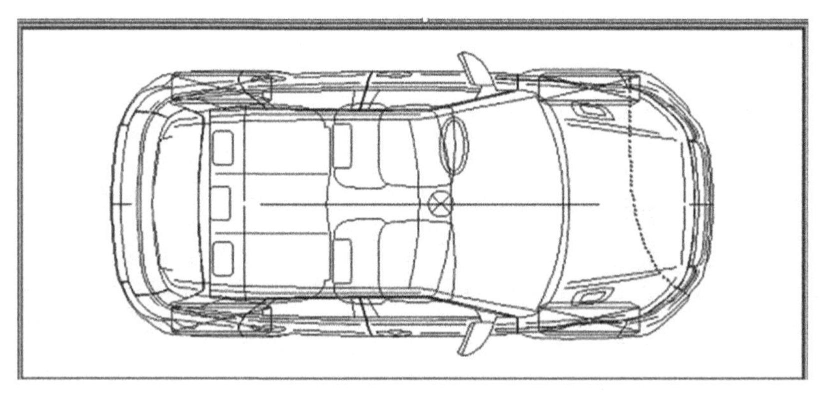

FIGURE 5.27 CAD image of the Range Rover vehicle, with the deformed profile drawn from the comparison with the crash test of Fig. 5.26.

The deformation is the classic one from a frontal impact at 40% offset, whose maximum depth is equal to about 41 cm and, from the Triangle method, a value of b_1 is obtained as follows:

$$b_1 = \frac{\text{EES}\,\gamma - b_0}{k\,C} = \frac{16.4 \cdot 1 - 2}{0.653 \cdot 0.41} = 52.5$$

The deformation of the object vehicle is schematically as shown in Fig. 5.28. The deformed profile of the side can be assimilated to a triangular deformation, whose maximum introflection is about 24 cm, the extension is

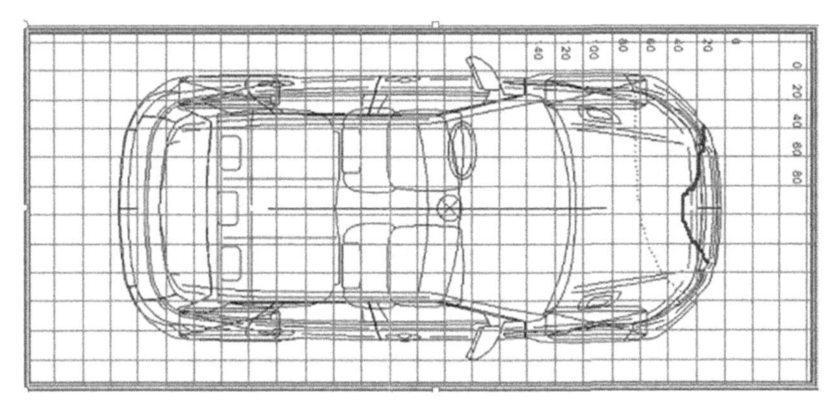

FIGURE 5.28 Representation of the Rover deformation starting from the available photographs.

equal to about 57% of the width of the front of the vehicle, and the PDOF is equal to about 25 degrees.

With these data, we get

$$\gamma = \sqrt{\frac{L_{100}}{L_d}\cos(\mathrm{PDOF})} = 1.26$$

and, with the b_1 already get, we obtain

$$\mathrm{EES}_O = \frac{1}{\gamma_O}[2 + b_1 k_O C_O] = \frac{1}{1.26}[2 + 52.5 \cdot 0.564 \cdot 0.24] = 5.74\,\mathrm{m/s} = 20.7\,\mathrm{km/h}$$

corresponding to a Rover energy loss of 31.1 kJ.

Energy equivalent speed evaluation by case comparison

For the Mercedes, from the "AutoExpert Hungary" database, it is possible to identify vehicles with similar damage in terms of magnitude and position to those in question, as shown in Fig. 5.29. The values of the deformation energy corresponding to the individual reference vehicles are distributed over a range of 10.7−17.8 kJ, with an average of 13.7 kJ. The vehicle that best approximates the deformation of the Mercedes could be the one with deformation energy of 17.8 kJ, by extension, shape, and stiffness class of the vehicle. All these values are lower than those calculated with the Triangle method, equal to 21.8 kJ. It is worth noting in this regard that the vehicles taken as reference are all older than the Mercedes vehicle in question and probably with a stiffness that is not entirely comparable. Therefore it is considered more reliable the evaluation made with the Triangle method, obtained starting from stiffness evaluated on the same type of vehicle.

For the Rover the reference vehicles from the "AutoExpert Hungary 2002" database, as shown in Fig. 5.30, can be used. The values of the

FIGURE 5.29 Examples of vehicles with side damages for a comparison with the damage of the Mercedes. From the top, clockwise, the following deformation energies are present: 15.4, 10.7, 12.9, 17.8, 13.1, and 12.1 kJ.

deformation energy corresponding to the individual reference vehicles are distributed more than a range of 20.2–42.9 kJ, with an average of 31.4 kJ. Despite being different vehicles from the Rover in question and with deformations that are not entirely comparable, the average value of the deformation energies is comparable with that obtained by the Triangle method.

5.2.6 Preimpact velocity calculation using the method based on momentum calculation

In the treated example, it is possible to estimate the initial speeds of the vehicles knowing the input and output directions and the output speeds, obtained from the analysis of the postimpact phase.

FIGURE 5.30 Examples of vehicles with front damage for comparison with Rover one. From the top, clockwise, the following deformation energies are present: 26.7, 42.9, 33.8, 30.7, 20.2, and 34.2 kJ.

Using Eq. (2.53):

$$m_1 V_1 \cos(\theta_1) + m_2 V_2 \cos(\theta_2) = m_1 \overline{V}_1 \cos(\overline{\theta}_1) + m_2 \overline{V}_2 \cos(\overline{\theta}_2)$$

$$m_1 V_1 \sin(\theta_1) + m_2 V_2 \sin(\theta_2) = m_1 \overline{V}_1 \sin(\overline{\theta}_1) + m_2 \overline{V}_2 \sin(\overline{\theta}_2)$$

with the values identified by the analysis of the postimpact phase and reported earlier, the vehicle preimpact speeds are obtained as follows:

Mercedes: $V = 36$ km/h
Rover: $V = 40$ km/h

To evaluate the postimpact angular speeds of the vehicles, we can apply Eq. (2.34):

$$m_1 y_1 \overline{V}_{1n} - m_1 y_1 V_{1n} - m_1 x_1 \overline{V}_{1t} + m_1 x_1 V_{1t} = J_1(\overline{\omega}_1 - \omega_1)$$

$$m_2 y_2 \overline{V}_{2n} - m_2 y_2 V_{2n} - m_2 x_2 \overline{V}_{2t} + m_2 x_2 V_{2t} = J_2(\overline{\omega}_2 - \omega_2)$$

assuming a negligible initial angular velocity ω for both vehicles and with the following values of x and y, obtained from Fig. 5.31:

	Mercedes	Rover
x	−0.87 m	1.73 m
y	−0.72 m	0.32 m
$\overline{V}_n = \overline{V}\cos\overline{\theta}$	−14.2 km/h	−22.4 km/h
$\overline{V}_t = \overline{V}\sin\overline{\theta}$	−32.0 km/h	−8.6 km/h

we get

Mercedes $\overline{\omega} = 1.9$ rad/s
Rover $\overline{\omega} = 2.5$ rad/s

A similar result can be obtained with a forward analysis, applying Eqs. (2.22) and (2.27), without the need to knowing the directions of the input and output speeds, but having to assume a value for the restitution coefficient.

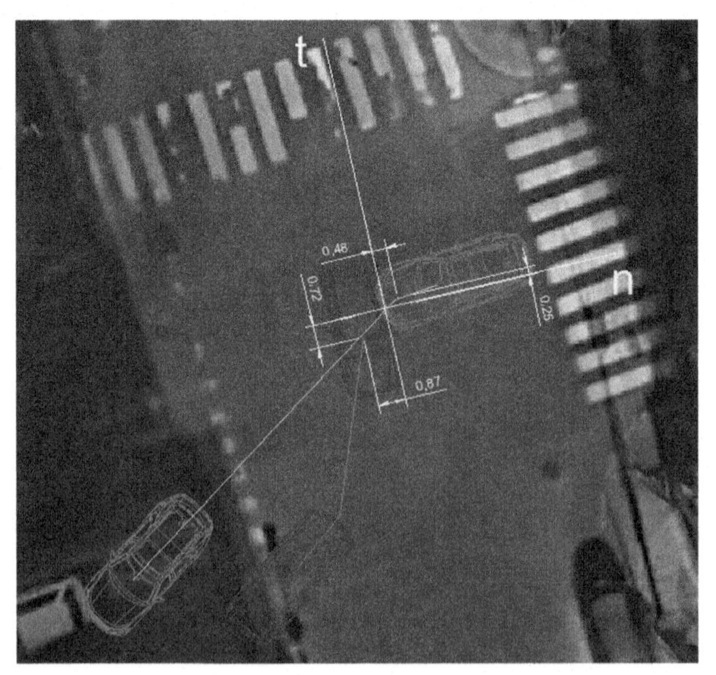

FIGURE 5.31 Position of the vehicles at the moment of impact on the road and quiet position, indicating the space covered.

Assuming as input speed the values that previously found, 36 and 40 km/h for Mercedes and Rover respectively, with zero initial angular speeds and $\varepsilon = 0$, from Eq. (2.27), we get

	Mercedes	Rover
I_n	−9110 N	9110 N
I_t	2948 N	−2948 N
PDOF	−23 degrees	72 degrees
$\overline{\omega}$	2.0 rad/s	2.6 rad/s
$\overline{\theta}$	245 degrees	202 degrees
μ	0.34	

The value of μ can still be considered acceptable because the vehicle deformations suggest that there has been no relative sliding in the tangential direction and that the opening of the friction cone is greater than $\tan^{-1}$ (0.34).

Applying Eq. (2.22), the final speeds are obtained as follows:

Mercedes: $V = 35$ km/h
Rover: $V = 24$ km/h

Results check

The results must be verified by comparing the value of the kinetic energy variation with that of the deformation energy estimated by the vehicle analysis.

Calculating the kinetic energy variation of the two vehicles between before and after the collision, we obtain

$$\Delta E_c = \frac{1}{2} m_1 \left(V_1^2 - \overline{V}_1^2 \right) + \frac{1}{2} m_2 \left(V_2^2 - \overline{V}_2^2 \right) - \frac{1}{2} J_1 \overline{\omega}_1^2 - \frac{1}{2} J_2 \overline{\omega}_2^2 = 63 \text{ kJ}$$

This value is higher than the deformation energy estimated by the vehicle analysis. The final angular speed of the Rover is higher than that of the Mercedes, despite having a lower arm of the resulting forces with respect to the center of gravity. The PDOF value obtained with the conservation of momentum, on the other hand, is comparable with that estimated based on vehicle deformations.

Since the calculation carried out with the conservation of momentum is sensitive to the input values (initial and final directions of the vehicles, post-impact speed, positions of the centers of gravity of the vehicles with respect to the reference system, centered on the center of impact and orientation of the contact plan), an analysis can be performed by simulating a number n times the impact, varying each time the input parameters within likely intervals, for example, with the Monte Carlo method. Among the n solutions found, therefore, only those that satisfy, within specific tolerances, some

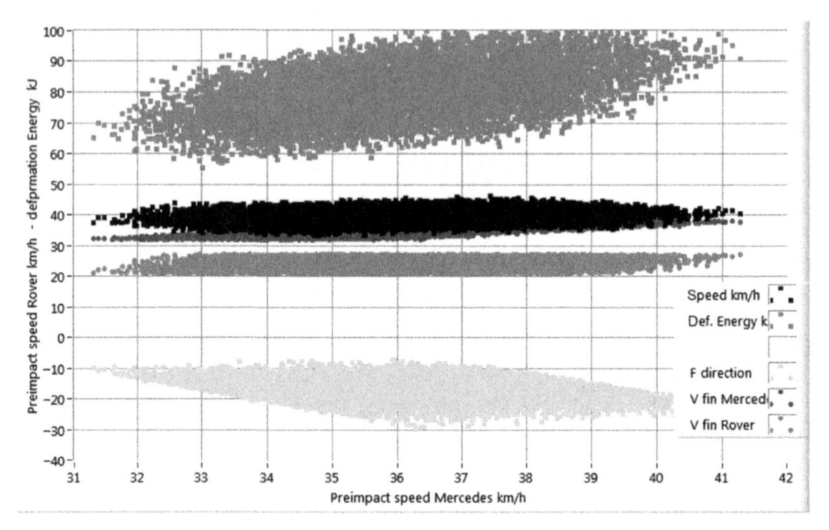

FIGURE 5.32 Results of the simulation with the Monte Carlo method: for each speed of the Mercedes vehicle (on the abscissa), the values of the Rover speed (on the ordinate) and the corresponding values of the deformation energy, the PDOF, and the final speeds of the vehicles are shown. *PDOF*, Principal direction of forces.

imposed constraints are considered. In the analysis that follows, the postimpact speeds of the vehicles, the direction of the PDOF (resulting from the contact forces between the vehicles) and the deformation energy have been assumed as constraints. The following table shows the values assumed for the analysis with the known directions and their plausible range of variation.

	Mercedes	Rover	Variation
M (kg)	2274	1891	± 20
$\overline{V}$ (km/h)	35	24	± 3
θ (degrees)	270	185	± 1
$\overline{\theta}$ (degrees)	246	201	± 3

The results of the analysis are shown in graphical form in Fig. 5.32. The lowest values of kinetic energy variation, compatible with the estimated deformation energy on vehicles, are obtained for initial Mercedes speeds of around 33 km/h and a corresponding speed value of around 40 km/h for the Rover.

5.2.7 Preimpact speed calculation based on deformations measurements

From the gyratory radii $k^2 = J/m$ and resultant forces h arms, we get the mass reduction factors (2.45): $\gamma = k^2/k^2 + h^2$:

Mercedes: 0.68
Rover: 0.69

and, applying Eq. (2.98):

$$\Delta V_1 = \frac{1}{m_1}\sqrt{2E_d m_c^*}$$

with deformation energy equal to that estimated by the analysis of the vehicles—53 kJ—and a zero restitution coefficient, the speed variations of the vehicles are obtained as follows:

Mercedes: ΔV 13.7 m/h
Rover: ΔV 16.5 km/h

Since the initial and final speed directions of the cars and the final speed module are known, as shown in the following table:

	Mercedes	Rover
M (kg)	2274	1891
$\overline{V}$ (km/h)	35	24
θ (degrees)	270	185
$\overline{\theta}$ (degrees)	246	201

The triangles of the velocities can be traced, as indicated in Fig. 5.33. Starting from the final velocity vector, known in module and direction and sense, a circle with a radius equal to the speed variation is drawn. The intersection of the circle with the direction of the initial speed determines the module of the initial velocity vector.

The following initial speeds result from the speed triangles:

Mercedes: 37.6 m/h
Rover: 31.7 km/h

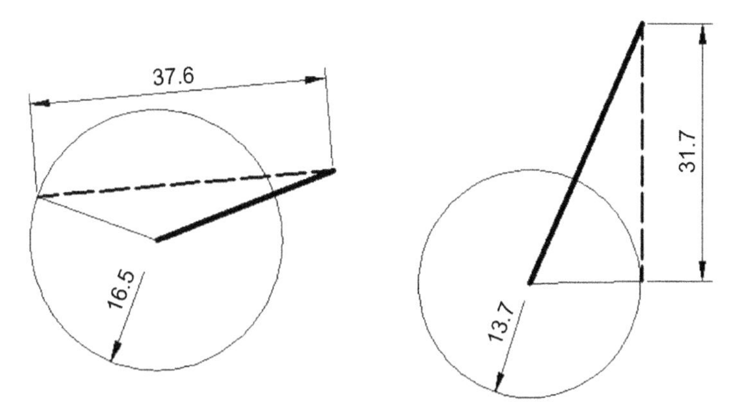

FIGURE 5.33 Speed triangles for Mercedes (left) and Rover (right). In continuous bold lines the final speeds are reported, in dashed lines the initial speeds and in continuous line the speed variations ΔV.

Results check

To evaluate the congruence of the results determined earlier with the deformation-based method, it should be verified that the speed triangles are such that the direction of the velocity variation vectors ΔV is the same for the two vehicles. Approaching each other the two triangles as in Fig. 5.34, we see that the two vectors are not perfectly aligned, even if this inconsistency is modest.

By slightly varying the directions of the final velocities, which is the most uncertain parameter, congruent solutions are obtained. For example, with a final direction variation of the Mercedes of 2 degrees, from 246 degrees to 248 degrees, we get the following triangle of speed for Mercedes, while remaining unchanged for Rover (Fig. 5.35).

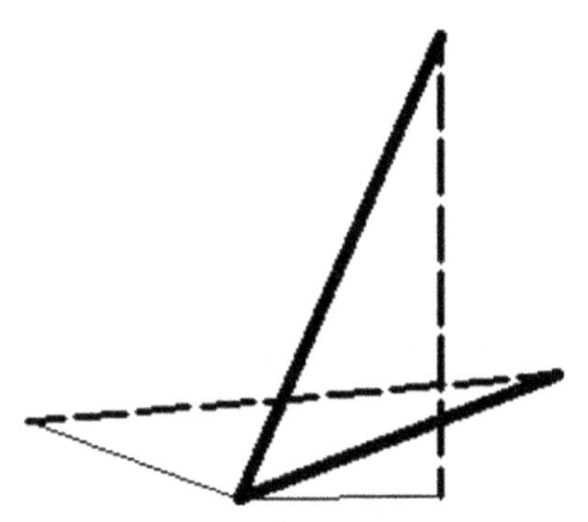

FIGURE 5.34 Approaching the speed triangles of the two cars to each other. It is noted that the velocity variation vectors ΔV, thin continuous lines, are not perfectly aligned with each other.

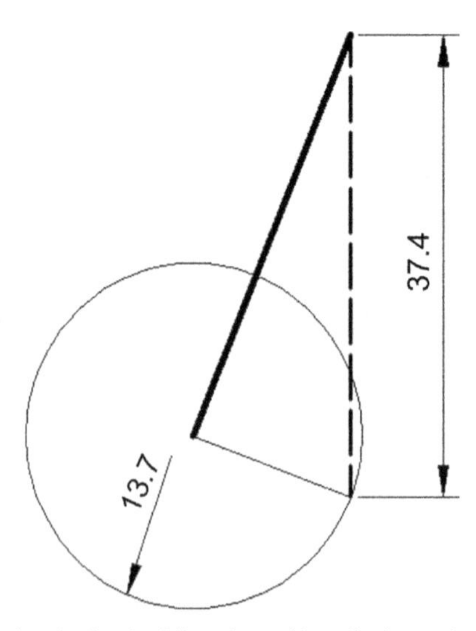

FIGURE 5.35 Speed triangle for the Mercedes, with a final speed angle of 248 degrees. In continuous bold lines the final speeds are reported, the dashed line shows the initial speed, and in continuous thin line the speed variation ΔV.

Approaching the speed triangles of the two vehicles each other, results in an alignment of the two vectors of speed variation ΔV, as shown in Fig. 5.36.

The resulting initial speeds are as follows:

Mercedes: 37.6 m/h
Rover: 37.4 km/h

5.2.8 Velocities from the video analysis

The analysis of the video allows determining with reasonable precision the vehicles speed immediately before the impact. Considering the moment of impact and another previous instant, as shown in Fig. 5.37, there is a displacement of the vehicle of about 6.96 m accomplished in seven frames (frame rate = 12 fps), equivalent to an average speed of 43.0 km/h.

Another speed estimate can be made considering only the last four frames, as shown in Fig. 5.38, in which the Rover travels a distance of

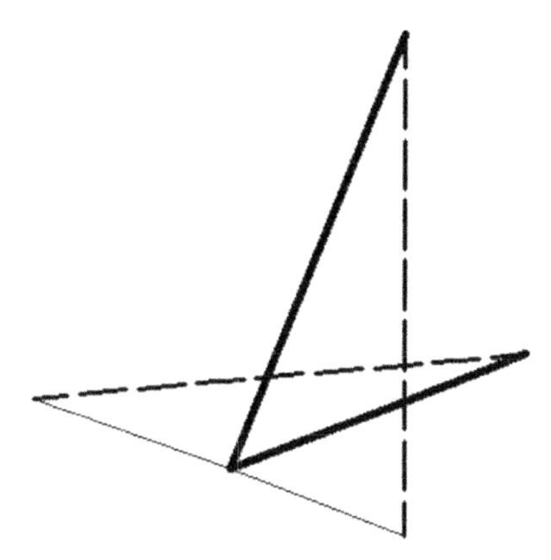

FIGURE 5.36 Approaching the speed triangles of the two cars to each other, the velocity variation vectors ΔV, thin continuous lines, are now aligned with each other.

FIGURE 5.37 Moments considered to evaluate the speed of the Rover at the moment of impact. The distance traveled is 6.96 m, accomplished in seven frames.

approximately 3.66 m in four frames, which corresponds to an average speed of 39.5 km/h. A similar evaluation carried out on the first three frames gives the result of 46.8 km/h. We can deduce that the vehicle was slightly slowing down and probably the speed at the moment of impact was around 39 km/h.

For Mercedes, making similar assessments between the instants shown in Fig. 5.39, in which the vehicle travels 3.25 m in four frames, we obtain an average speed of 35.1 km/h.

In conclusion, the following table summarizes the preimpact velocity values obtained with the various methods, which show an excellent congruence:

Speed (km/h)	Mercedes	Rover
Video analysis	35.1	39.0
Conservation of momentum	36.0	40.0
Monte Carlo simulation	33.0	40.0
Deformation measurements	37.6	37.4

FIGURE 5.38 Moments considered to evaluate the speed of the Rover at the moment of impact. The distance traveled is 4.2 m, accomplished in four frames.

FIGURE 5.39 Moments considered to evaluate the speed of the Mercedes just before the impact. The distance traveled is 3.25 m, accomplished in four frames.

References

Brach, R., Brach, R., 2001. Vehicle Analysis and Reconstruction Methods. SAE International, Warrendale, PA.

Brach, R., Brach, R., 2005. Vehicle Accident and Reconstruction Methods. SAE International, Warrendale, PA, ISBN: 0-7680-0776-3.

Gillespie, T.D., 1992. Fundamentals of Vehicle Dynamics. SAE International, Warrendale, PA.

Han, I., 2015. Impulse—momentum based analysis of vehicle collision accident using Monte Carlo simulation methods. Int. J. Automot. Technol. 16 (2), 253–270.

Ishikawa, H. 1985. Computer simulation of automobile collision—reconstruction of accidents. In: Paper 851729, 29th Stapp Car Crash Conference. Washington, DC.

Ishikawa, H., 1993. Impact model for accident reconstruction-normal and tangential restitution coefficients. In: SAE Paper No. 930654.

Ishikawa, H. 1994. Impact center and restitution coefficients for accident reconstruction. In: SAE Paper No. 940564.

Kolk, H., Tomasch, E., Sinz, W., Bakker, J.J, Dobberstein, J., 2016, Evaluation of a momentum based impact model and application in an effectivity study considering junction accidents. In: International Conference "ESAR—Expert Symposium on Accident Research. Hannover, Germany.

Macmillan, R.H., 1983. Dynamics of Vehicle Collisions. Inderscience Enterprises Ltd, La Motte Chambers.

Pacejka, H.B., 2005. Tire and Vehicle Dynamics. SAE International, Warrendale, PA.

Smith, R.A., Noga, J.T., 1982. Accuracy and sensitivity of CRASH. In: SAE Technical Paper 821169.

Sutherland, I.E., Hodgman, G.W., 1974. Reentrant polygon clipping. Commun. ACM 17, 32–42.

Vangi, D., 2008. Accident Reconstruction. Firenze University Press, ISBN: 978-88-8453-783-6 (in Italian).

Vangi, D., 2009. Simplified method for evaluating energy loss in vehicle collisions. Accid. Anal. Prev. 41, 633–641.

Vangi, D., 2013. A simplified model for analysis of post-impact motion of vehicles. Proc. Inst. Mech. Eng., D: J. Automob. Eng. 227 (6), 39–48.

Index

Note: Page numbers followed by "*f*" and "*t*" refer to figures and tables, respectively.

K

L

M

N

O

P

T

U

V

W

Printed in the United States
By Bookmasters